STRUCTURE AND EVOLUTION OF THE CONTINENTAL LITHOSPHERE

Editors

HENRY N. POLLACK
Dept. of Geological Sciences, University of Michigan

and

V. RAMA MURTHY
Institute of Technology, University of Minnesota

PERGAMON PRESS

OXFORD · NEW YORK · TORONTO · SYDNEY · PARIS · FRANKFURT

U.K.	Pergamon Press Ltd., Headington Hill Hall, Oxford OX3 0BW, England
U.S.A.	Pergamon Press Inc., Maxwell House, Fairview Park, Elmsford, New York 10523, U.S.A.
CANADA	Pergamon Press Canada Ltd., Suite 104, 150 Consumers Road, Willowdale, Ontario M2J 1P9, Canada
AUSTRALIA	Pergamon Press (Aust.) Pty. Ltd., P.O. Box 544, Potts Point, N.S.W. 2011, Australia
FRANCE	Pergamon Press SARL, 24 rue des Ecoles, 75240 Paris, Cedex 05, France
FEDERAL REPUBLIC OF GERMANY	Pergamon Press GmbH, Hammerweg 6, D-6242 Kronberg-Taunus, Federal Republic of Germany

This edition 1984

Library of Congress Catalog Card No: 56-4104

British Library Cataloguing in Publication Data

Structure and evolution of the continental lithosphere.—
(Physics and chemistry of the earth; v. 15)
1. Earth—Crust
I. Pollack, H. N. II. Murthy, V. Rama
III. Series
551.1'3 QE511

ISBN 0 08 030023 5

Published as Volume 15 of the journal *Physics and Chemistry of the Earth* and supplied to subscribers as part of their subscription. Also available to non-subscribers.

Printed in Great Britain by A. Wheaton & Co. Ltd., Exeter

CONTENTS

PREFACE

If the 1970's can be seen as a decade of intense study of the oceanic lithosphere, then the 1980's will likely be viewed as the decade of investigation of the continental lithosphere. The life cycle of oceanic lithosphere we have come to recognize as relatively simple: it is created by emplacement of magmatic material at oceanic ridges, moves laterally across the ocean basins losing heat, contracting and becoming more dense along the way, and ultimately is returned to the earth's interior to be reheated and reprocessed. The continental lithosphere by contrast is more diverse petrologically, displays a wide variety of both compressional and extensional structures, and has a more complex, lengthy and perhaps irreversible history. More and more interest is being turned to a multifaceted study of the continental lithosphere, in part motivated by purely scientific concerns, and in part by the resources resident in the lithosphere.

This volume of PHYSICS AND CHEMISTRY OF THE EARTH comprises four papers that focus on the geochemical, structural, thermal and seismological characteristics of the continental lithosphere. Sacks and Snoke review the seismological observations that define the thickness and internal structure of the continental lithosphere. Weaver and Tarney assemble geochemical and petrological arguments to estimate the composition of the continental crust, which in turn constrains both the process and timing of continental growth. Kröner discusses petrological and structural processes active in the cores of ancient mobile belts, with implications for the growth and stabilization of lithosphere. Finally Morgan presents a thorough discussion of the thermal structure and evolution of the continental lithosphere, including an in depth analysis of the global heat flow data set. Taken together these four papers offer a broad perspective of the geochemical and geophysical character of the continental lithosphere and of the processes that have created and shaped it and by which it has evolved through time.

Henry N. Pollack
V. Rama Murthy
Editors

SEISMOLOGICAL DETERMINATIONS OF THE SUBCRUSTAL CONTINENTAL LITHOSPHERE

I. S. Sacks* and J. A. Snoke†

**Department of Terrestrial Magnetism, Carnegie Institution of Washington, Washington, DC 20015, U.S.A.*

†Virginia Tech Seismological Observatory, Department of Geological Sciences, Virginia Polytechnic Institute and State University, Blacksburg, VA 24061, U.S.A.

ABSTRACT

The distinction between lithospheric and asthenospheric parts of the suboceanic upper mantle is well understood. The velocity and anelasticity of the asthenosphere are readily determined to be less than that of the overlying lithosphere. We therefore use similar criteria to define the less well understood continental lithosphere. A simple conductive cooling model is consistent with seismological results for oceanic regions, but the generation of continental plates does not appear to be related to any single process. Stable shields, aseismic platforms, rift zones or basins, mountain regions and subduction zones all have different deep structure, and there is no simple model which explains trends in thickness - even within a given region. Various seismological techniques have been used to determine the structure of the continental lithosphere: these include refraction profiles, surface waves, anelasticity, body waves and travel-time anomalies. In a few cases similar results come from independent techniques, but in others the estimates are in controversy. A few generaliztions can be made: Basins, rift valleys and alpine regions all have resasonablu well developed low velocity zones (LVZ) beginning at depths less than 100 km. Aseismic platforms may have LVZ's, but in some regions (e.g., eastern Australia) the implications from different studies are inconsistent. Results from Precambrian shield regions indicate that subcrustal velocities are high, and that if an LVZ exists, it is at most a moderate one starting at a depth possibly much greater than 100 km. Japan has an anomalous structure; the subcrustal velocites are low and show no decrease with depth, but a Q reversal at about 65 km may indicate the base of the continental lithosphere. The results for shield regions and for Japan raise questions concerning the validity of the assumption that the LVZ always represents the asthenosphere.

1. INTRODUCTION

In this chapter we discuss the subcrustal continental lithosphere from a seismological perspective. The goal is to determine the properties of the continent and its root which translate as a tectonic unit. This is not a comprehensive review, but instead a selection from a range of studies to cover the various aspects. Anticipating our conclusions - seismology has not yet been successful in defining the continental lithosphere in all areas. This situation may be expected to improve in the future as the seismological techniques become more refined and more constraints from other disciplines become incorporated.

The distinction between lithospheric and asthenospheric parts of the suboceanic upper mantle is well understood. We therefore use the same criteria to define the less well understood continental lithosphere. The oceanic crust is typically about 6 km thick and overlies material with a shear-wave velocity, S_n, of about 4.6 km/sec and compressional-wave velocity, P_n, velocity of greater than 8 km/sec. There is a velocity reversal in both V_P and V_S beginning some tens of kilometers below the moho, with V_S decreasing to about 4.3 km/sec in most areas and V_P decreasing by a smaller proportion. The depth to the top of the low-velocity zone (LVZ) increases from about 30 km near the mid-oceanic ridges to more than 80 km at distances from the ridges corresponding to ages in excess of 80 million years. Except in the immediate vicinity of the ridges, the depth to the LVZ increases as the square root of the age suggesting that conductive cooling might be the controlling parameter. It has also been observed that the material in the LVZ has higher anelasticity (lower Q) than the rest of the upper mantle; Q values in the hundreds compared to Q values in the thousands. A common assumption is that the material in the LVZ, being hotter, has a small degree of partial melt but is otherwise similar to the higher-velocity material above it.

The assumption is made that this thermal boundary is also the mechanical (plate) boundary, so that the top of the LVZ is the top of the asthenosphere. Based on this oceanic model, we tentatively define the subcrustal lithosphere as the high velocity, high Q "lid" in the upper mantle overlying a lower velocity, low-Q region.

Unlike the oceanic case, the generation of continental plates does not appear to be related to any single process such as conductive cooling. Stable shields (e.g., South America, eastern North America, the Baltic Shield, North-Central Africa, western Australia), aseismic platforms (e.g., eastern Australia, southern US), rift zones or basins (e.g., east African rift, the Great Basin in the western US), mountain regions (e.g., the Alps) and subduction zones (e.g., western South America, Japan) all have different deep structures, and there is no simple model that explains trends in thicknesses, even within a given region.

In the next section we describe the techniques which have been used to determine the velocity structure, the anelasticity structure and the thickness of the subcrustal continental lithosphere. Paralleling the oceanic lithosphere structure, the lithosphere-asthenosphere boundary in a given region corresponds to the most shallow *significant* velocity reversal or Q reversal.

2. TECHNIQUES FOR DETERMINING SUBCRUSTAL LITHOSPHERIC STRUCTURE

Refraction Profiles: Much of our knowledge about the *P*-wave velocity distribution with depth in the crust and upper mantle comes from refraction studies. Such studies allow the determination of the seismic velocity-depth structure from observed travel times of various seismic arrivals for varying distances from the (earthquake or explosion) source. Such data also provided the first evidence for a velocity reversal beneath a continent.

Seismic waves from explosions or earthquakes are recorded by an approximately linear array of seismographs. The time-domain traces for these signals are superimposed in a travel-time vs. distance plot as shown in Fig. 1. By comparing waveforms on successive observing-station traces, one can interpret "branches" of travel-time curves for the more distinct seismic arrivals.

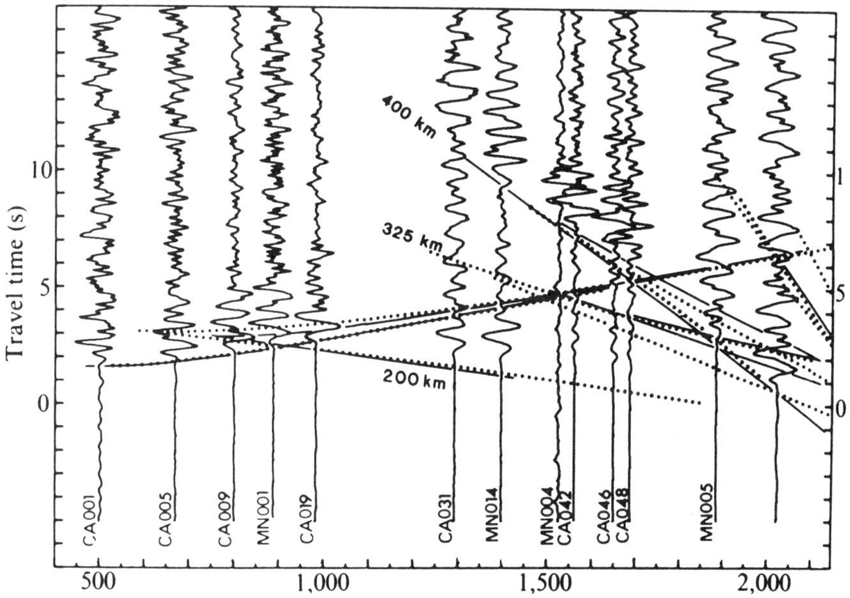

Figure 1: Vertical-component record section of an event from the Banda Sea arc recorded at seismograph stations in central Australia (Hales, et al., 1980). Travel-time curves are included for their model 8 (dotted line) and Leven's CAPRI model (solid line). The models are shown in Fig. 2. The branch labeled 200 km corresponds to rays refracted below the 200 km discontinuity. This branch displays a rapid decrease in amplitude beyond an epicentral distance of 1,400 km, corresponding to the onset of a negative velocity gradient beneath the 200 km discontinuity. (After Leven, et al., 1981. Reprinted by permission from *Nature*, Macmillan Journals Limited.)

The (multibranched) travel-time curve can be inverted to produce a velocity-vs-depth structure (e.g., Fig. 2). For each travel-time curve there is a unique velocity model, but, as is seen from these figures, it may be possible to construct more than one travel time curve for a given set of time

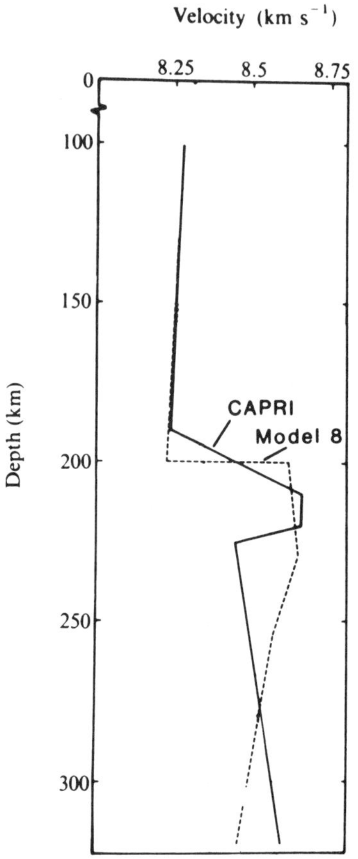

Figure 2: The velocity-depth models corresponding to the travel-time curves in Fig. 1. The structure in model *CAPRI* is required to match the waveform data. This model is unusually complex. (After Leven, et al., 1981. Reprinted by permission from *Nature*, Macmillan Journals Limited.)

traces. (High signal-to-noise ratios and close station spacing help reduce such ambiguity.) The structure is further refined - especially regarding the depth resolution of the velocity structure - if there are seismograms for the same region from multiple sources at different depths. (Of course, there will still be uncertainties on the order of the earthquake depth-location error.)

Downgoing seismic rays traveling through a region in which the seismic velocities increase rapidly with depth result in multiple arrivals at some distance. This, in turn, can result in a "triplication" on a travel-time curve (Fig. 3a). If first-arrival data only are used, the resulting travel-time curve is two intersecting straight lines with different slopes. In the simplest case, these slopes give the velocities of the two layers. Tucker, et al., (1968) produced a map (Fig. 4) of observed P_n velocities - the mantle P velocity just below the moho - beneath the continental United States. There is a correlation of the P_n velocities with tectonic region: higher values beneath the northeastern shield regions, lower values beneath the western Basin and Range region (which is often interpreted as a rift zone).

The inclusion of secondary arrivals, as is shown in Fig. 1, can be used to distinguish between sharp discontinuities and steep gradients in the velocity-depth structure. Most refraction profiles have considered only *P* waves because of the relatively weak excitation of shear waves by explosion sources. There are a few refraction studies, however, which lead to subcrustal *S* velocities (e.g., Hales, Muirhead and Ryan, 1980a; Mayer-Rosa and Mueller, 1973).

If there is a rapid decrease in velocity with depth, the resulting travel-time curve may have a "shadow zone", which is an epicentral distance range with no strong first arrivals. Such a zone also results in an offset in the travel-time curve and/or a reduction in the amplitudes of the arrivals in some distance range (Fig 3b). The presence of such a LVZ in the upper mantle was first suggested by Gutenberg (1926) based on the observed rapid decrease of *P* wave amplitudes with increasing epicentral distance up to about 1600 km (from Pasadena). Beyond this distance the *P*-wave amplitudes increased by more than a factor of 10. Subsequently Gutenberg (1948) found that the low velocity region appeared better developed for *S* waves than for *P* waves. Gutenberg's LVZ extends from about 80 km to about 230 km depth.

In general, the observation distances for a refraction profile should be more than five times the depth to be probed: To observe a shadow zone corresponding to a LVZ at a depth greater than, say, 150 km would thus require a line extending considerably more than 750 km over a region which is reasonably homogeneous laterally over that range. Over such distances in most continental regions there are heterogeneities which are of the same order as the resolution of the refraction method. In general, a dense network of recording stations is required to obtain the most reliable solutions. Only a few such regional studies have been carried out. Some which have found LVZ's at depth are Muirhead, et al., (1977) who find a LVZ for *P* in the depth range 155-190 km in southeastern Australia using large, timed explosions whose signals were recorded at seismic stations in the Western Plains of New South Wales; Hales, et al.,(1980b) whose refraction profile across northern Australia suggests a possible LVZ for *P* in the depth range 200-325 km (see Figs. 1 and 2 above); and Yegorkin and Pavlenkova (1981) who find a LVZ for *P* beneath the East Siberian platform in the depth range 200-240 km based on long-range observations. Masse and Alexander (1974) reported a 20 km thick LVZ for *P* centered at about 100 km beneath Scandinavia and western Russia based on Russian explosions recorded at NORSAR, &ar.100 km apperature seismic array in southern Norway. However, the lack of seismic coverage in the distance range 320-1200 km in their study means that they had no direct evidence pertaining to the upper mantle structure in 60-200 km depth range. A more recent study by King and Calcagnile (1976), which determined the *P*-wave distribution beneath Fennoscandia and western Russia using data from NORSAR of presumed explosions in continental Russia,

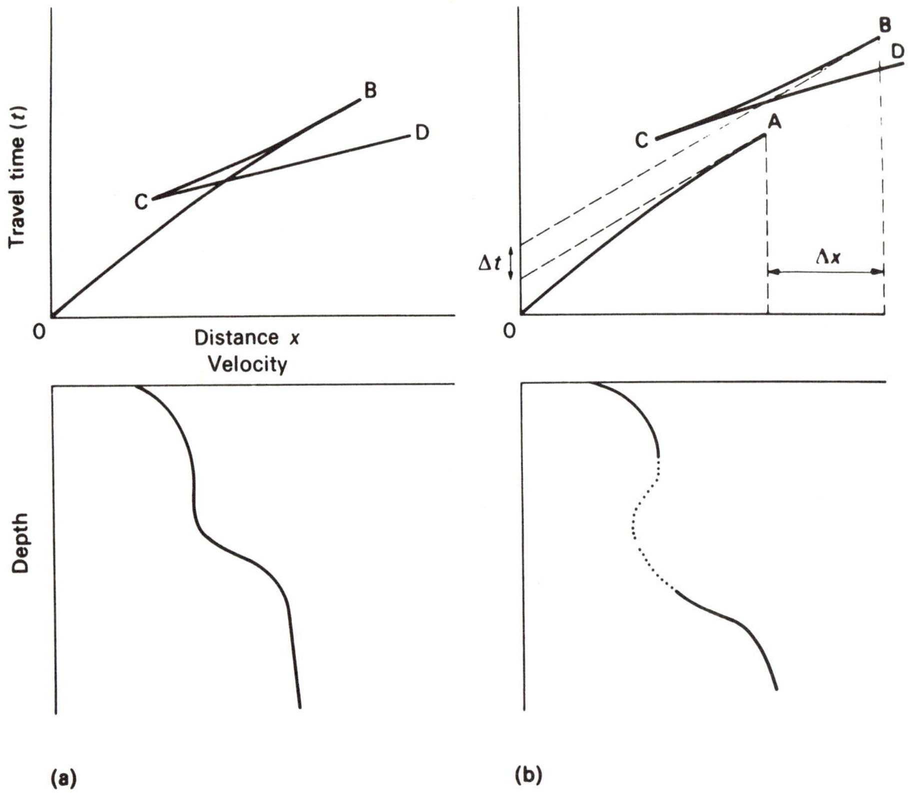

Figure 3: Time-distance curves for continuous variation of velocity with depth, showing *(a)* triplication with retrograde branch *BC* caused by a steep, positive velocity-depth gradient, and *(b)* the effect of a LVZ causing a gap in arrivals between *A* and *B*, offsets Δx and Δt between *A* and *B* and the retrograde branch *BC* from the steep, positive velocity-depth gradient below the LVZ. (From Bott, 1982, who adapted it from an article by Giese, 1976, *Explosion Seismology in Central Europe*, p. 132, Springer-Verlag.)

included some coverage towards the upper end of that distance range and found no evidence for an LVZ.

In recent years *synthetic seismograms* have been used in studies of refraction profiles. In this approach, the source time history is assumed or solved for separately, and comparisons between the observed and calculated waveforms are used with iteration and/or inversion to determine the velocity and Q structures along the travel path between the energy source and a set of stations. When applied to modeling refraction profiles, synthetic seismograms have the potential of providing better resolution because they incorporate directly amplitude information as well as arrival times. Burdick (1981) (see also Helmberger (1973) and Burdick and Helmberger (1978)) has derived a compressional velocity profile for the tectonically active western part of the US by matching observed long-period waveforms with synthetic seismograms. His model has a maximum lid velocity of 8.05 km/sec and an LVZ which starts at 65 km depth and which has a minimum velocity of 7.70 km/sec at 103 km.

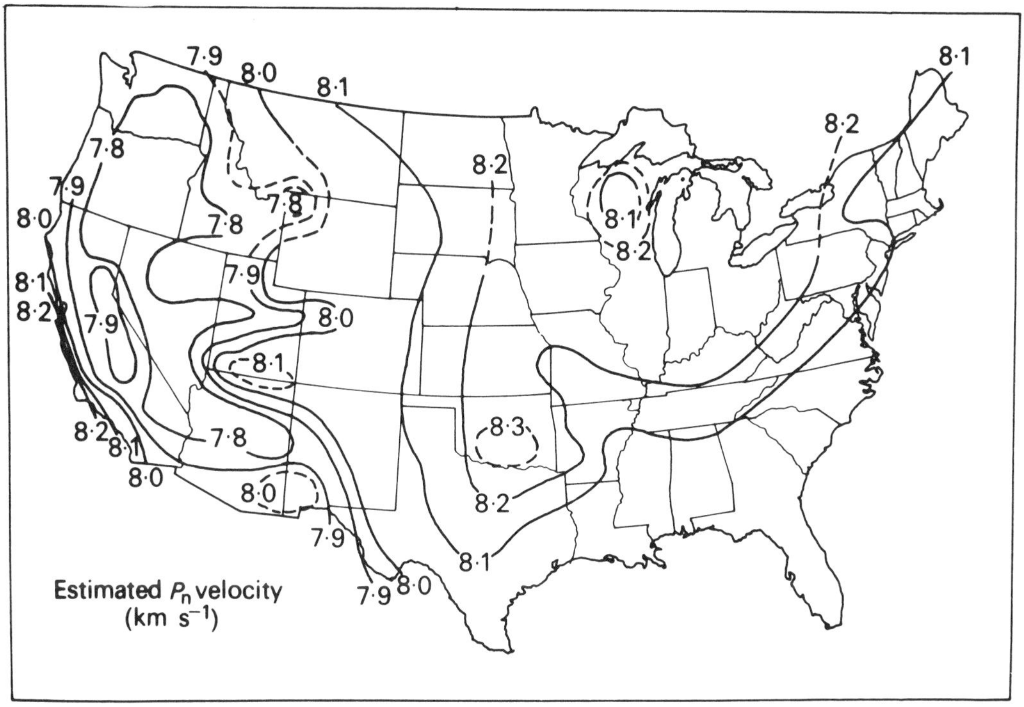

Figure 4: Estimated P_n velocity in the United States. Note that in the platform and shield regions in the east, velocities are higher, whereas in the Basin and Range region in the west, the velocities are lower. (After Tucker, et al., 1968.)

Burdick also develops models for central and eastern US (his "stable continent" model S8), and Given and Helmberger (1980) developed a model for central and southern Europe (their "northeast Eurasia" model K8). The only difference between these models for depths less than 250 km is that model S8 has no LVZ, and model K8 has one between 140 km and 230 km depth (velocities decreasing from 8.34 km/sec at 140 km depth to a minimum of 8.04 km/sec at 170 km).

Surface Waves: Surface wave studies provide the primary constraints on the *S*-wave velocity distribution and on the presence of a LVZ from among the various techniques. Thus, they have provided, in turn, much information on lithospheric thickness. Surface waves are long-period (10-150 second) seismic waves which travel along the Earth's surface. There are two kinds of surface waves, Rayleigh waves which have an elliptical particle motion in the vertical plane containing the source and receiver, and Love waves which have a horizontal particle motion perpendicular to that plane. The particle motion for these waves as a function of depth is shown in Fig. 5. The vertical scale in these figures is approximately proportional to wavelength; that is, longer wavelengths sample greater depths. Surface waves are effectively guided waves, which means that there are different solutions (called *modes*) to the eigenvalue problem to determine wave number as a function of frequncy.

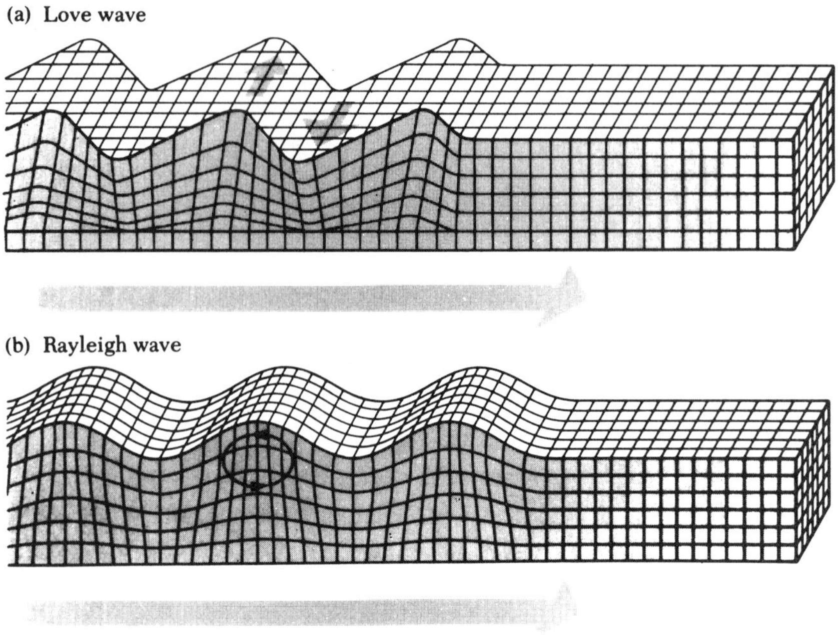

Figure 5: Ground motion at the earth's surface: Love surface waves and Rayleigh surface waves. Note that the deformation decreases with increasing depth. (From Bolt, *Nuclear Explosions and Earthquakes*, W. H. Freeman and Company, 1976. All rights reserved.)

The propagation speed for a surface wave depends on the medium properties within approximately a wavelength of the surface. As the seismic velocities generally increase with increasing depth, the propagation speed increases with increasing wavelength. Thus, at large epicentral distance there will be pronounced dispersion in the surface waves with the longer periods arriving before the shorter periods. There are two methods which have been used to specify this frequency-dependent dispersion: the *group velocity*, the velocity with which the energy is transmitted, and the *phase velocity*, the velocity of individual peaks and troughs. Group velocity determinations require only observations at a single station, while phase velocity determinations - employing the fewest assumptions - utilize observations from two stations. The arrival-time-vs-period data are reduced to group and/or phase velocity spectra which are then compared with model-derived curves. Rayleigh-wave dispersion curves depend on the depth dependence of the P velocity, S velocity and density, while the Love-wave dispersion curves depend on only the S velocity and density. However, even for Rayleigh waves, the dispersion curve shapes are most affected by the S-wave velocity. Hence, surface waves studies complement refraction studies in the sense that the latter are most effective in providing information about, and constraints for, the P -wave velocity structure in a region.

There is an intrinsic nonuniqueness in the inversion of phase or group velocity curves due to the fact that one is trying to obtain information about an earth model which has more degrees of freedom than are available in the data. Hence surface-wave studies utilize as much other independent data as possible - e.g., refraction data - to constrain the models.

Dispersion curves are often similar in shape for different regions with similar tectonic settings, and there are usually characteristic differences between such curves from regions which are geologically dissimilar. Figure 6 shows a family of observed dispersion curves for different tectonic regions. Note the pronounced differences in shapes in the individual curves over the period range 40 to 120 seconds.

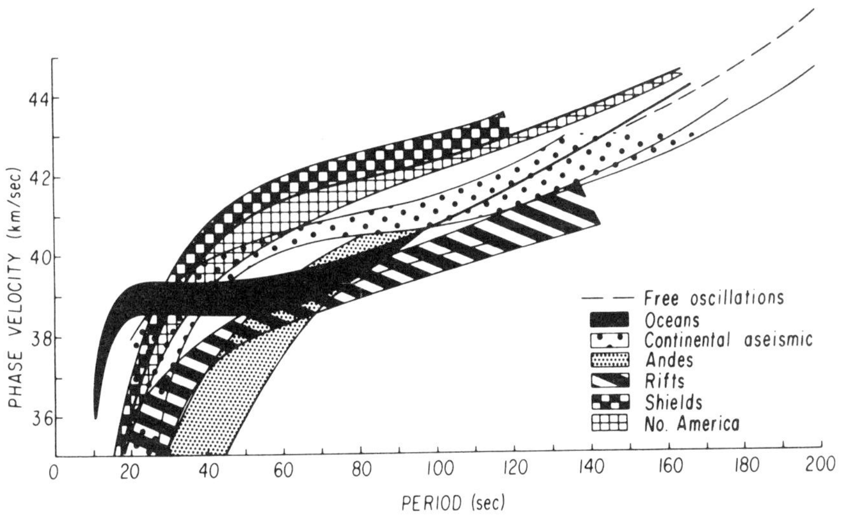

Figure 6: Comparison of the span of Rayleigh-wave phase velocities for shields, aseismic platforms, rifts, oceans and mountains. Only for the oceanic case does the shape of the curve truly *require* a LVZ (the flat portion from 20 to 60 seconds). The determination of a LVZ in other regions requires additional data. (After Knopoff, 1972.)

Figure 7 shows three V_S models illustrating the range in subcontinental structures obtained by inverting fundamental-mode Rayleigh phase velocity curves such as those shown in Fig. 6. Shield regions typically have high lid velocities and have either no resolvable LVZ or one which is at best only moderately developed (a decrease of about 0.2 km/sec). The shields which have a LVZ, exhibit characteristacally S_n (submoho) velocities which are ~4.7 km/sec and a channel velocity of ~4.5 km/sec (e.g., the Canadian shield), while shields for which an LVZ is not required to match the dispersion, typically have velocities which are ~4.5 km/sec from the moho to about 200 km depth (Cara, et al., 1980 for the Baltic shield and Russian platform).

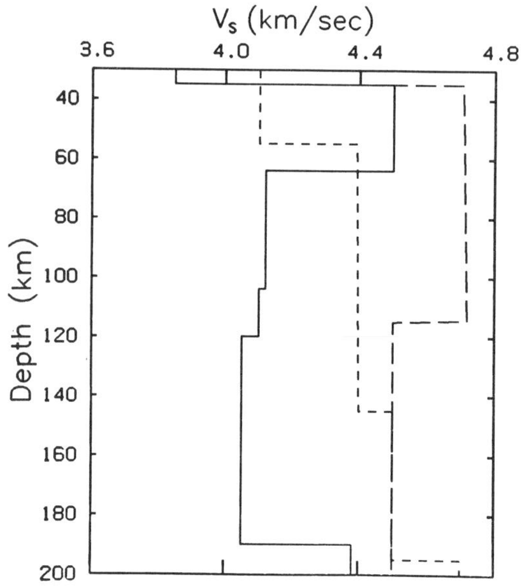

Figure 7: Span in continental-region velocity-depth structures inferred from Rayleigh-wave fundamental-mode phase velocity observations: the Great Basin (solid lines) after Priestley and Brune, 1978; the Canadian shield (long dashes) after Brune and Dorman, 1963; eastern Japan (short dashes) after Shiono, et al., 1980.

Other regions, such as eastern Japan (Fig. 7), or the Andes beneath central Peru (James, 1971), have low or moderate lid velocities (~4.5 km/sec) and no resolvable LVZ. Some regions, such as the Great Basin in the western US (Fig. 7), have a thin, moderate velocity lid over a well developed LVZ.

Dispersion curves imply models which represent an averaging over the entire path between the epicenter and station in the single-station methods or between the two stations in the two-station method. Surface waves, therefore, have a limited resolution in regions where the geologic structure changes laterally in a rapid manner. However, several research groups have developed methods to deal with lateral heterogeneities along paths; e.g., Panza, et al., (1980) have developed a method of phase-velocity regionalization and have applied it to determine the gross features of the lithosphere-asthenosphere structure beneath all of Europe. Their results for Europe are shown in Fig. 8.

Resolution studies (e.g., Knopoff and Chang, 1977; Der, et al., 1970; Wiggins, 1972; Nolet, 1977) indicate that layer thicknesses in the upper 300 km cannot be resolved to better than within 30 km and 0.7 km/sec respectively. As an example of this point, in a study of the central US, Mitchell and Herrmann (1979) note that their data are not capable of resolving a LVZ, and that if one were claimed it would simply reflect assumptions in their starting model. Also, Knopoff and Schlue (1972) note that, on the basis of their data for the East African rift, it is not possible to choose between a structure

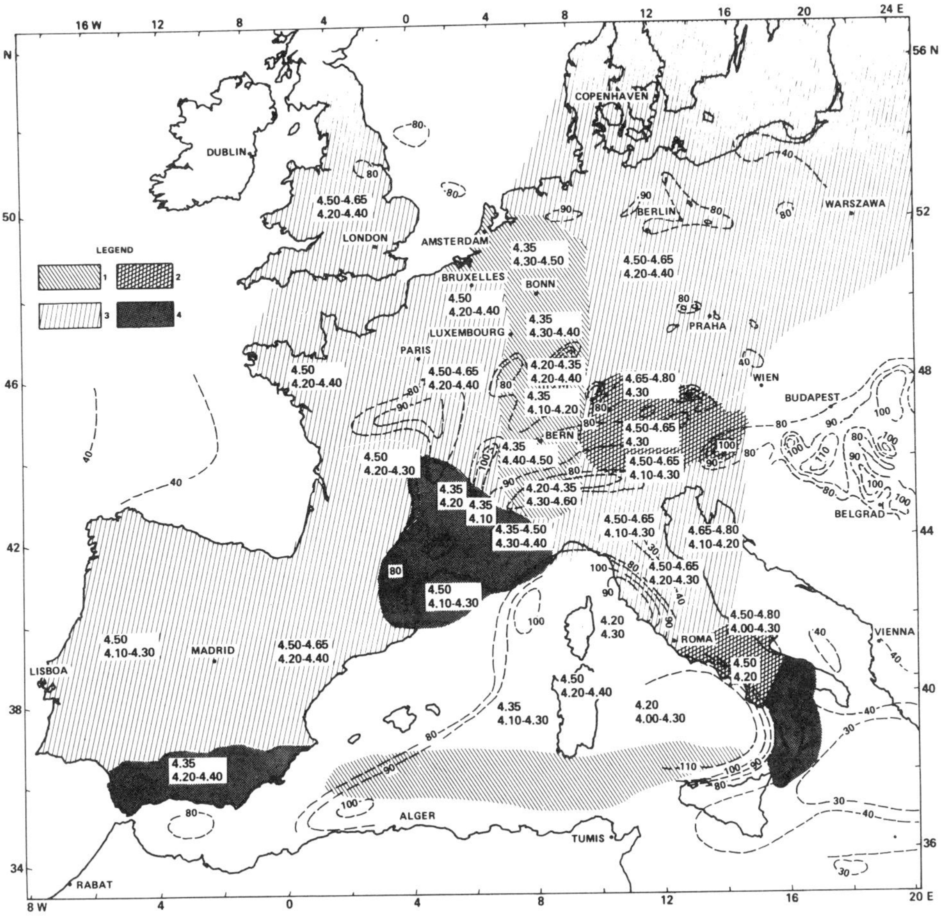

Figure 8: Schematic representation of the upper-mantle properties obtained by the "hedgehog" inversion procedure (e.g., Keilis-Borok and Yamovskaja, 1967). Different hatchings correspond to the maximum possible "lid" thickness *h*: *1 = h < 45 km, 2 = h < 75 km, 3 = h < 105 km, 4 = h < 135 km.* Numbers in the upper line represent the S_n velocity and italic numbers (lower line) gives the range of *S*-wave velocities (km/sec) in the asthenosphere. The broken isolines mark the surface het flow anomalies for Q ≤ 40 mW/m² and Q ≥ 80 mW/m². (Redrawn from Panza, et al., 1980.)

such as the Great Basin model shown in Fig. 7 or one in which the velocity is a constant 4.25 km/sec down to a depth of 200 km. Again, as with the shields, the existence of a LVZ depends most directly on whether or not there is an assumed or required higher lid velocity.

Clark and Stuart (1981) measured Rayleigh-wave fundamental-mode phase velocities for two paths in Great Britain: one from northern Scotland parallel to the Caledonides to southwest Ireland, and one from northern Scotland to southern England through Precambrian basement. For the Caledonides path they found a well developed LVZ with velocities of 4.2-4.4 km/sec beneath a lithosphere which is 110-190 km thick with an S_n velocity of 4.5-4.6 km/sec. Along the other path the lithosphere has comparable velocities and is at least 140 km thick with a LVZ that is less pronounced or absent (V_S > 4.4 km/sec).

Patton (1980) has studied the crust and upper mantle structure of the Eurasian continent using a continent-wide Rayleigh-wave phase-velocity regionalization involving a data set consisting of dispersion curves for 42 paths in Eurasia. He subdivided Eurasia into five provinces. The provinces with their corresponding S_n velocity, depth to the LVZ and channel velocity are as follows: Indian shield (4.72 km/sec, 118 km, 4.45 km/sec); northern platforms and shields (4.72 km/sec, 73 km, 4.55 km/sec); forelands and coastal plains (4.5 km/sec, 73 km, 4.4 km/sec); tectonic (4.55 km/sec, 83 km, 4.3 km/sec) and plateau (4.45 km/sec, 125 km, 4.40 km/sec). In no provinces were there any reversals in density or V_P. Of particular interest is the fact that his plateau velocity model is quite similar to James' (1971) model for the altiplano in Peru (with little or no LVZ), and his Indian shield model is similar to Brune and Dorman's (1963) model for the Canadian shield (Fig. 7) except that the Indian shield model has a 0.1 km/sec lower channel velocity, which Patton felt was required to match the phase velocities better over the period range of 50-90 sec.

Using higher-mode Rayleigh-wave phase-velocity data gives a considerable improvement in resolution in structure at depth over using only the fundamental mode (e.g., Nolet (1977, our Fig. 9). This follows from the fact that

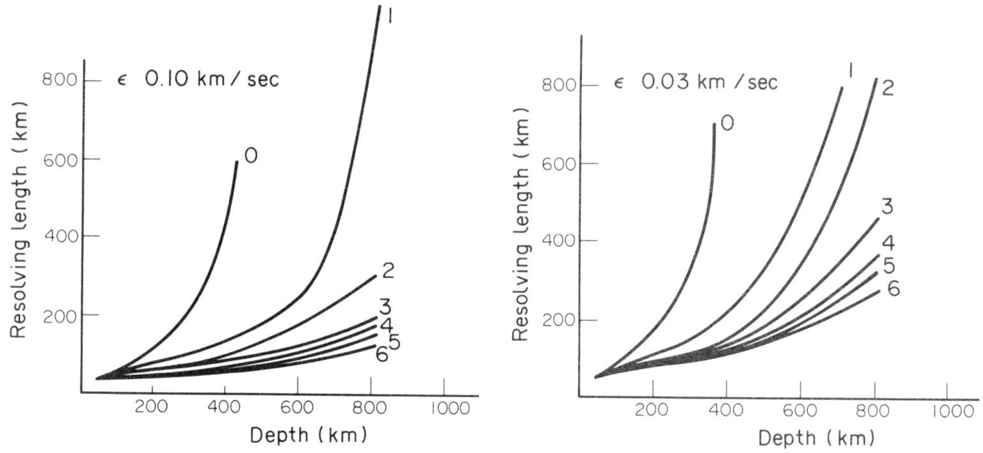

Figure 9: Reduction of the resolving length for the shear-wave velocity with the addition of higher Rayleigh modes shown as a function of depth for two velues of the model error ε. Numbers denote the highest mode number present in the data set. Note that the resolving power for deeper structure improves substantially if higher modes are included. (After Nolet, 1977.)

for a given period, higher mode waves sample greater depths than does the fundamental mode. In addition, using more than one mode decreases the nonuniqueness in matching the models. A problem with higher-mode data is that the dispersion curves for different modes may overlap so that it is difficult to separate the individual modes. Several authors have succeeded in isolating individual Rayleigh-wave higher modes by application of stacking procedures on data from linear arrays of long-period stations: e.g., Nolet (1977) for western Europe and Cara (1979) for several regions within the US. In Cara's study, the western US was found to have a low lid velocity (7.8 km/sec) and a broad, well-developed LVZ centered at about 140 km; in the northeasetern and central US he finds a lid velocity of about 4.7 km/sec and a slight LVZ (minimum velocity of 4.5 km/sec at 220 km) which was not required to fit only the fundamental-mode data. For paths across western Europe, Nolet finds *S* velocities in the lid of 4.4-4.5 km/sec above a LVZ centered at about 180 km depth with a minimum velocity of about 4.33 km/sec. His data do not rule out an additional, shallower LVZ between 70 and 120 km, as was reported by Mayer-Rosa and Mueller (1973). However, Nolet comments that it is probably possible to fit their (Mayer-Rosa and Mueller, 1973) body-wave data with a simple velocity increase near 100 km depth. Nolet also finds a slight density reversal beginning at about 200 km depth. However, in a later study using a subset of these data, Cara, et al. (1980) do not find a low density zone. The Cara, et al., study also used high-mode Rayleigh waves to study northern Eurasia. They find the velocity to be a rather constant 4.45-4.55 km/sec from the moho to a depth of about 250 km - no thick high-velocity lid and no LVZ. Priestley, et al., (1979) refined the Priestley and Brune (1978) Great Basin model using synthetics to model observed *Sa* phases, which they interpret as a superposition of several Rayleigh higher modes. In particular they verify the 4.5 km/sec lid velocity (Fig. 7).

In summary, surface waves have been used to verify the depth range and approximate velocity contrast of an LVZ if it is well developed and at depths shallower than about 150 km. Surface wave studies cannot rule out thin LVZ's although - particularly when higher-mode data are available - they can pose constraints on the average velocity over a given depth range. Finally, surface wave studies give little information concerning the abruptness of a transition region at the top or bottom of a LVZ: that is, whether it takes place over a few kilometers or a few tens of kilometers.

Anelasticity: The amplitude on a seismogram of a seismic wave depends on a number of factors apart from the strength of the source (earthquake or explosion). Examples of such factors are the types of wave and velocity structure through which it travels. Propagation is not loss-free, and the imperfect elasticity of the earth causes preferential attenuation of higher frequencies. Hence, waves will show more high-frequency loss if the raypaths are predominately in high anelasticity (low-Q) material. Beneath the oceans, the lithosphere has a very high Q, (probably many thousands) and overlies an athenosphere with far higher attenuation (the compressional-wave Q, Q_P, in the hundreds). The actual value of about 1000 for the average continental lithosphere Q is not particularly diagnostic: the heterogeneity (on all scales) of the continental lithosphere causes significant energy loss through non-anelastic mechanisms such as scattering. Continental lithosphere Q values are more variable and lower than those found beneath the oceans. However, the observed values are still well above any expected asthenospheric value (typically in the low hundreds).

The large contrast between lithospheric and asthenospheric Q values make Q studies particularly useful in determining the thickness of the lithosphere. Q structure in a region can be determined using either body waves or surface waves. The same limitations and assumptions would apply as those involved with the determination of velocity structure using the same waves.

Body waves: By choosing (body-wave) ray paths which penetrate to different depths, one can - in principle - determine the Q structure for the region. Q is normally determined from the high-frequency slope of displacement-amplitude spectra and, most stably, from spectral amplitude ratios that minimize source effects, crustal-structure filters and other non-anelasticity-related effects. To minimize the effects of structure on determination of the spectra, the widest possible frequency range needs to be used. Sacks and Okada (1974) have utilized such spectral techniques in a study which maps and contrasts the Q_P structure in the subduction zones beneath Japan and western South America. In these regions there is a favorable distribution of seismicity which allows Q determinations with minimum assumptions. An example is shown in Fig. 10, which shows the ray paths studied to determine the Q structure beneath Japan. Ray paths wholly in the lithosphere have a high apparent Q, whereas those partially in the asthenosphere have a significantly lower Q value. In the case of a continent (Japan) adjacent to a subduction zone with its attendant Benioff zone, the paths from earthquakes at increasing depths will eventually penetrate beneath the lithosphere. The depth at which the spectral slope increases (i.e. apparent Q decereases), indicates the thickness of the continental lithosphere. In the case of continents which have not developed a high velocity lid, the anelasticity structure may be the most effective measure of lithosphere thickness. Sacks (1975) determined the spectra of *P* and *S* waves at the broad frequency range seismograph at Matsushiro in western Honshu, Japan, from earthquakes on the subduction zone, (see Fig. 10a) at depths ranging from shallow (about 30 km) to nearly 300 km. The apparent Q_P for earthquakes down to about 60 km depth was 1190, a normal value for the continental lithosphere. For paths starting below 77 km, however, the apparent Q_P was only 315, indicating a substantially asthenospheric path. The interpretation is that the lithosphere thickness is between depths of 60 and 77 km. (Of course there is some uncertainty in the earthquake depths and this possible error affects the thickness estimate.) It was further reported that the Q_P values in the asthenosphere wedge below continental Japan and above the subduction zone are about 200 - similar to values found for oceanic asthenosphere. The resulting Q structure for Japan is shown in Fig. 11. Further anelasticity studies, similar to the above, are

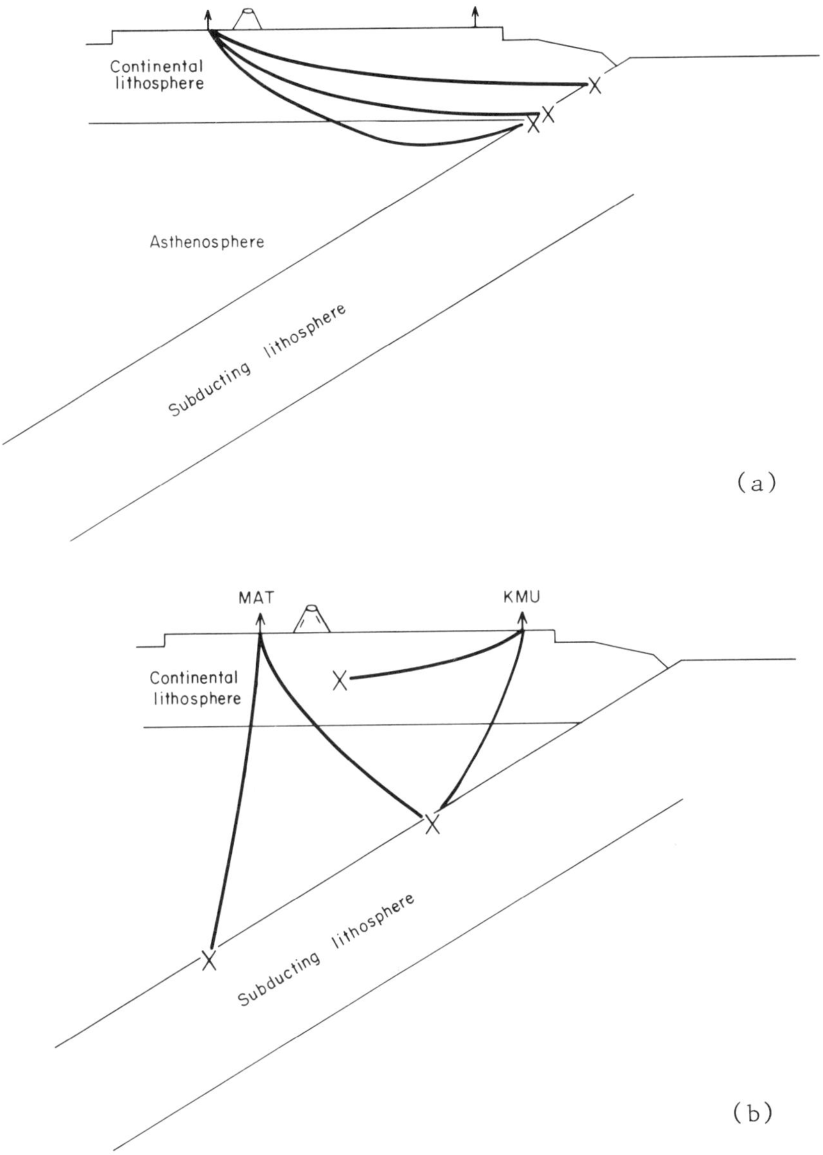

Figure 10: *(a)* Schematic of subduction zone showing raypaths studied to determine Q of the lithosphere. Crosses indicate earthquakes. There is a significant difference in the apparent Q between the two lower paths. *(b)* Raypaths studied to determine Q of the asthenosphere. Crosses indicate earthquakes. The two paths from the shallower earthquake in the subducting plate gave higher Q to KMU than to MAT (after correcting for the effect of the continental lithosphere). The deep event gave the same Q value to MAT as the shallow event did. (After Sacks, 1975.)

needed to confirm these results for Japan and should be performed in other regions as well.

Helmberger (1973) has calculated a Q map of the region beneath western US using comparisons between time-domain seismogram traces recorded at US stations from large nuclear explosions at the Nevada Test Site and synthetic seismograms generated for different assumed velocity structures. Figure 12 shows the relative amplitude between the (infinite Q) synthetics and the

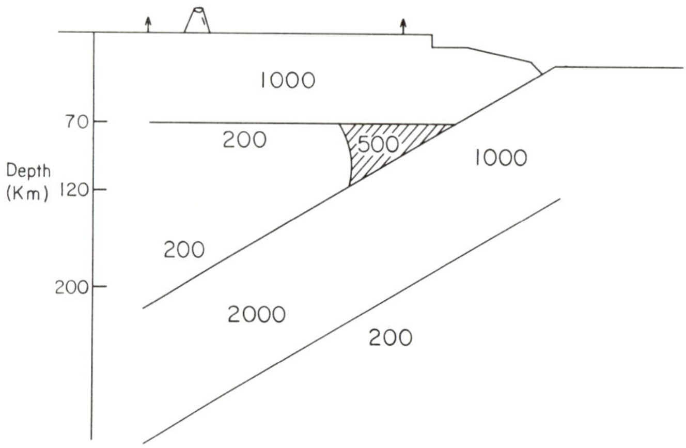

Figure 11: Q model for the Japan region. The hatched region has an anomalously high Q for asthenosphere, $Q_P = 500$. Normal oceanic lithosphere has $Q_P = 200$. In an earlier study, before the existence of the anomalous region was recognized, Sacks and Okada (1974) suggested a value of 400 for a homogeneous asthenosphere. (After Sacks, 1975.)

(normalized) observations plotted as a function of the epicentral distance. Assuming a low Q - LVZ coincidence, he was able to match the observed waveforms with a model which has a low-Q region and LVZ from about 70 km to 120 km. Basically the same structure was found in a later study by Burdick and Helmberger (1978), which examined alternative approaches of handling the attenuation operator when deriving synthetic seismograms.

Surface waves: Surface waves of different periods have a different depth distribution of wave energy. It is therefore possible to determine the Q structure of the path from the (source corrected) spectrum of the arrival. Lee and Solomon (1975) have inverted for Q using both Rayleigh and Love wave data for paths in western and east-central US. Identifying the base of the lithosphere as the depth at which Q decreases sharply, they obtain lithospheric thicknesses of 80±20 km and 130±30 km respectively for the two regions. They also find that an increase in Q with depth within the lithosphere is resolvable, a result found as well by Sacks and Okada (1974) in their body-wave Q-study of subducting plates.

In the first of two subsequent papers, Lee and Solomon (1978) invert their Love-wave western North American data set simultaneously for phase velocity and attenuation. In that study they used their 1975 results for velocity and Q as a starter model. While their final results were quite similar to the initial models, they felt that the Q structure was better resolved. In the second followup study, Lee and Solomon (1979) extended their simultaneous-inversion technique to Rayleigh waves and applied it to oceanic paths, paths in the western US and paths in east-central North America. The results are again similar to those of their 1975 study. They note that Q is lower in the low Q zone beneath the western US than it is beneath the central Pacific.

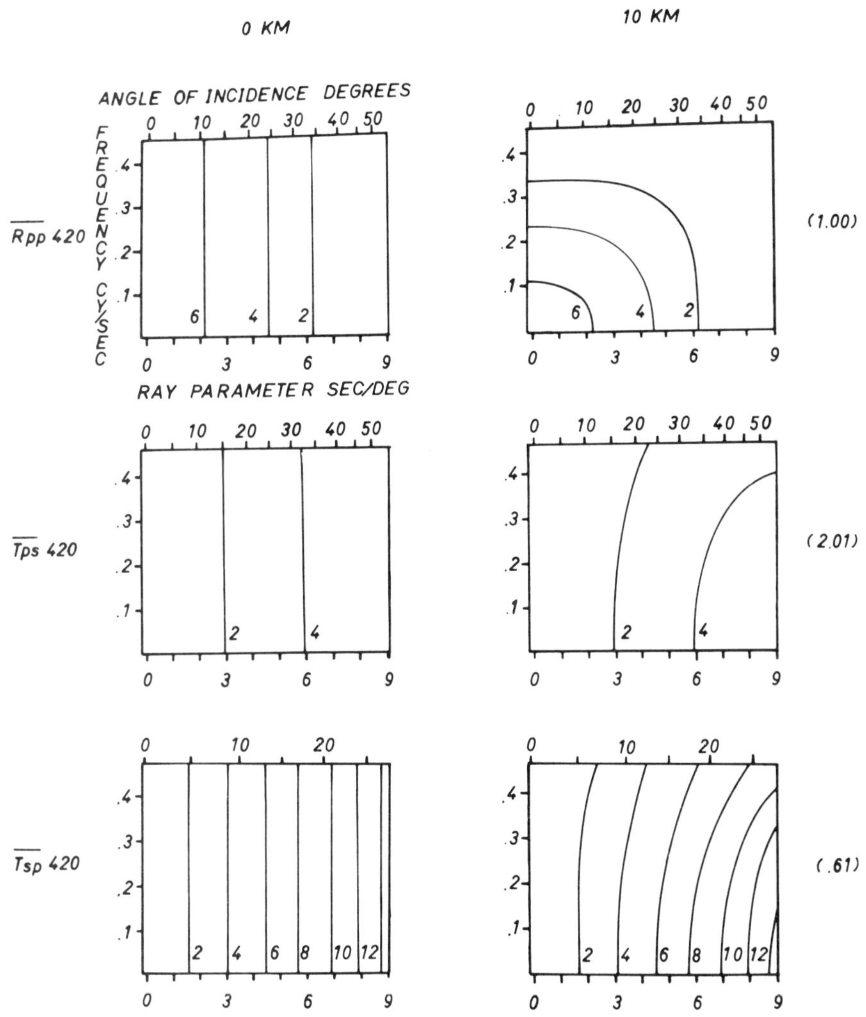

Figure 12: Effects of transition-sone thickness on the coefficients for *P-P* reflection and for *P-S* and *S-P* conversion on refraction at a discontinuity with a velocity contrast thought to exist at the 420-km discontinuity. The *P* and *S* velocities are given respectively by 8.75 km/sec and 4.67 km/sec above the discontinuity and 9.41 km/sec and 5.32 km/sec below the discontinuity. These curves are for a ray going from higher to lower velocities. The critical parameter is the magnitude of the velocity contrasts; the results would be quite similar if this discontinuity were at a different depth and/or if the sign of the velocity contrast were reversed. (After Ward, 1978.)

Burton and Bennell (1976; see also Burton, 1977) note that the spectra for Rayleigh waves are characteristically different between oceanic paths and paths which are primarily continental. The decrease with depth of Rayleigh-wave amplitudes depends on wavelength; the amplitude of a Rayleigh wave at a depth of one wavelength (into a homogeneous halfspace) is about 20% of its surface ampltidue. Burton and Bennell therefore suggest that the period at which inferred Q decreases rapidly can be used to derive a depth of the start of the low-Q region. They used nuclear blasts in Novaya Zemlya and in the South Sinkiang Province in China as the energy source. Thus, provided that the explosion does not release much tectonic stress, the radiated spectrum is simple, and amplitude variations can be ascribed to the path. For their continental and oceanic paths, such changes occur at 0.03 Hz and 0.045 Hz, which implies depths of 125 km and 70 km for the continental and oceanic lithosphere respectively.

In his study of the Eurasian crust and upper mantle, Patton (1980) solved for Rayleigh-wave Q in the different provinces. As he notes, a finer regionalization of Q is required before it is possible to make detailed comparisons between Q structure and velocity structure. However, he was able to resolve significant differences in Q over the periods of 25-60 seconds in three regions: for stable northern platforms and tectonic provinces east and west of central Asia.

Interface Determinations - Converted Phases: Under optimal conditions *converted phases,* (seismic energy which changes from shear-wave motion to compressional-wave motion, or vice versa, at an interface at depth) can provide detailed information about the location and sharpness of the boundary at the bottom of the lithosphere.

If there is enough change in the elastic properties over a depth range which is small in extent compared to the wavelength of the incident energy, there will be reflected and converted waves as well as refracted waves. Such waves can be either longitudinally or transversely polarized. For reflected or converted phases to be observable, the velocity contrast must be at least several percent over a depth range of at most several kilometers, and the incident angle of the rays to the discontinuity must be in a (restricted) range leading to efficient energy conversion (Fig. 13). Such phases may produce

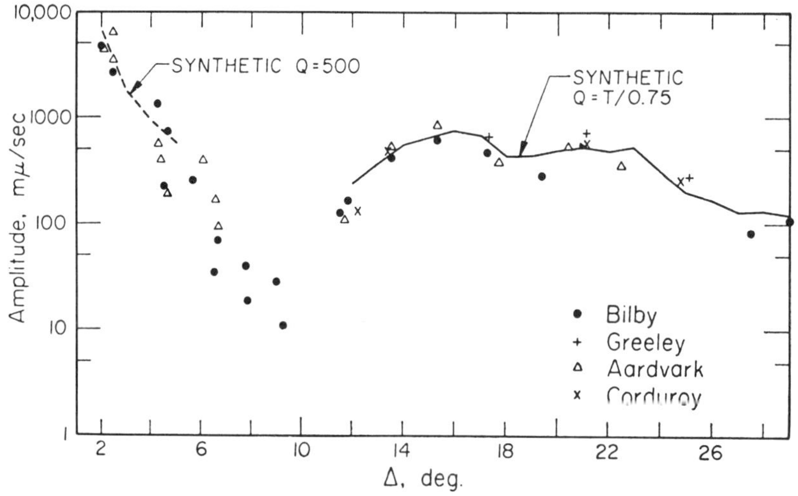

Figure 13: Observations (symbols), normalized to the HNME station values, and calculated amplitudes including Q. The explosions are all located at NTS. (After Helmberger, 1973.)

distinct arrivals at seismographs, but only for certain source-interface-station geometries will the *interpretation* of such arrivals be unambiguous. From the differential travel time between the converted-phase arrival and the direct arrivals, the depth to the conversion region can be determined. From the amplitude ratio between the direct and converted phase and an assumed source mechanism, the velocity contrast can be derived (although the scatter in the resulting estimate may be high).

For seismograms which are broadband in frequency, comparisons of the frequency content of the two arrivals allow constraints to be placed on the extent of the transition region at conversion (Fig. 13). Knowing the focal mechanism of the earthquake source, one can use the polarity of the converted phase to ascertain whether the velocity increases or decreases with depth across the discontinuity.

Two recent studies, (Sacks and Snoke, 1977; Sacks, et al., 1979) found evidence, based on converted phases, for velocity reversals beneath continental shield regions. These velocity reversals are interpreted as being the lithosphere-asthenosphere boundary. The Sacks and Snoke (1977) study interpreted arrivals in western South America from nearby, deep-focus earthquakes to be *sp* or *ps* arrivals which converted at an approximately horizontal interface at a depth of 400±35 km. Figure 14 shows the ray path geometry,

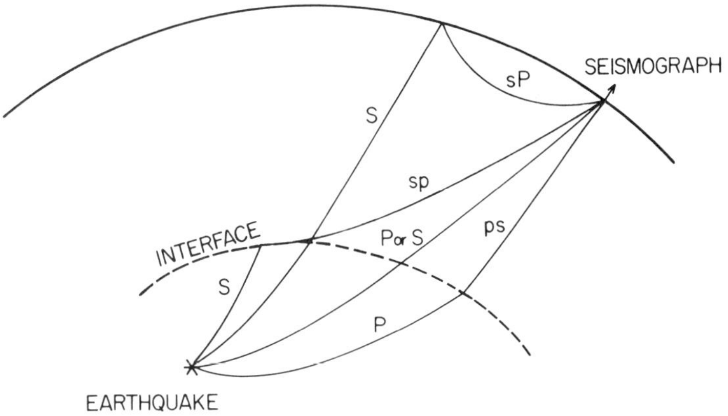

Figure 14: Raypaths of the phases considered in the South American converted phases study. (After Sacks and Snoke, 1977.)

Fig. 15 shows the station and epicenter locations and Fig. 16 some samples of the data. Sacks, et al., (1979) interpreted precursors to *S* at NORSAR from large earthquakes in the epicentral range 70-82° range as *Sp* arrivals converting at an approximately horizontal interface at a depth of 235±15 km beneath the Baltic shield. Fig. 17 shows the ray geometry used in that study and Fig. 18 depicts the array-conversion-point geometry and sample waveforms. The 1978 study differs from the 1977 one in that the data in the later study are from longer period waves and array-processing techniques are used in identifying and interpreting the converted phases.

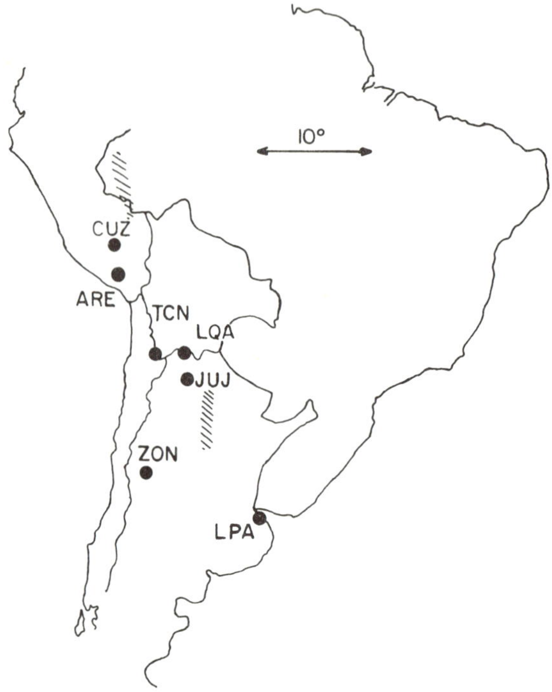

Figure 15: Location map showing seismograph stations and the epicentral regions (hatched zones) of analyzed earthquakes. CUZ and TCN are Carnegie broad-band stations, LPA and ARE are WWSSN stations, and the rest are mechanical seismographs. (After Sacks and Snoke, 1977.)

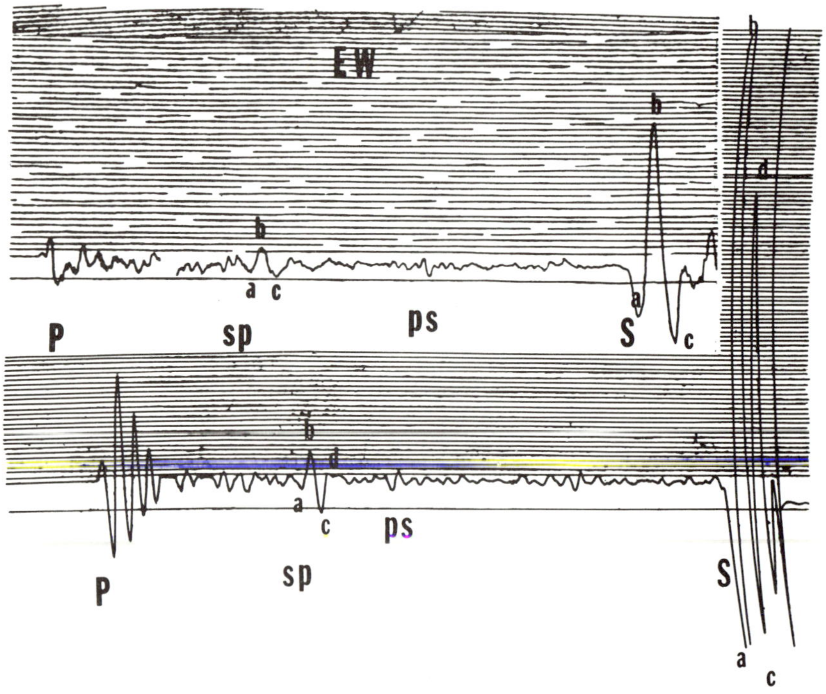

Figure 16: Photographs of EW component seismograms from ZON for two events. The lower-case letters indicate the correlation of the peaks and troughsin the *sp* and *S* waveforms. (After Sacks and Snoke, 1977.)

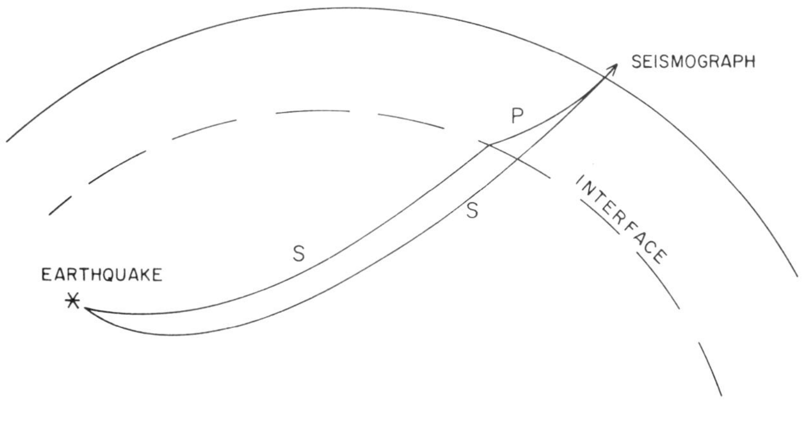

Figure 17: Raypaths for direct *S* and *Sp*. (After Sacks, et al., 1979.)

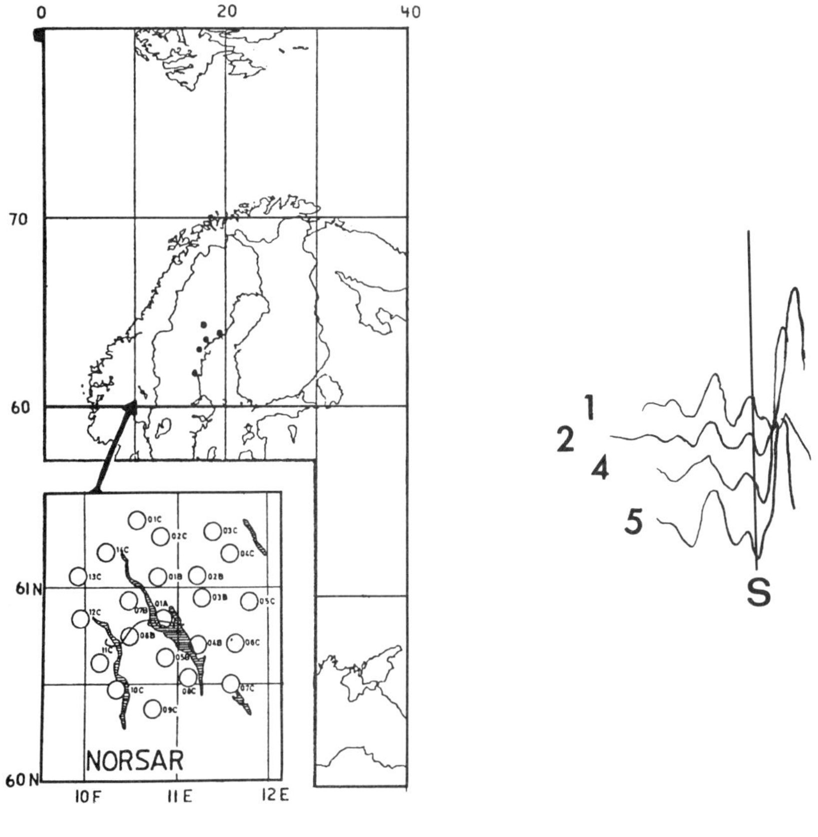

Figure 18: Map of the region around the NORSAR seismic array with data indicating calculated conversion points, assuming a conversion depth of 235 km for the *Sp* phases used in the study. Sample vertical-component waveforms for four of the events analyzed are shown on the right. These are alligned on the *S*-arrival time as determined from the transverse component. (After Sacks, et al., 1979.)

Thea above studies, taken alone, simply identify a velocity reve~sal of at least 5% in V_S over a transition region of at most several kilometers. Supplementary evidence is, of course, required to associate this interface with the lithosphere-asthenosphere boundary.

For the Sacks and Snoke South American converted phase study it is the polarity of these arrivals - and the sharpness of the interface - which distinguishes those phases from phases resulting from conversions at the normal "~400-km" discontinuity. This latter discontinuity represents a fairly rapid *increase* of velocity with depth and has been most commonly observed by multiple branching of travel-time curves, and is thought to be associated with the olivine-spinel transition in the mantle (e.g., Anderson, 1967). Using array processing of *Ps* converted-phase arrivals with periods greater than 10 seconds observed at NORSAR, Vinnik (1977) inferred an interface at about 400 km depth, and Baumgardt (1981) found a rapid velocity increase at about 450 km based on arrivals he interpreted as *Sp* conversions beneath the LASA array in north-central US. In both of these studies this boundary is interpreted as the normal 400-km discontinuity.

The crust and upper mantle beneath western Oregon and Washington states have been studied by Langston and Blum (1977), using synthetics to model simultaneously source parameters and velocity structure, and by Langston (1977), using converted *Sp* and *Ps* phases. These studies found the moho at about 15 km and a very well-defined LVZ bottoming at about 55 km beneath Puget Sound and 45 km beneath Corvallis, Oregon. Even though the two study regions are only a few hundred kilometers apart, the inferred shapes of the LVZ were quite different. Langston (1979) suggests that these differences might be caused in part by the fact that the layering is not horizontal in these regions, yet the modeling assumes horizontal layers. Because of this possible uncertainty, we do not include these studies in our Tables. Langston (1979) also reports arrivals which he interprets to be *Sp* converting at about 145 km depth with a relative polarity to *S* that suggests a velocity increase with depth. He suggests that this might be the bottom of the LVZ.

Travel-time anomalies: Aki, et al., (1976, 1977), developed a scheme for mapping three-dimensional velocity structure using travel-time anomalies measured across an array. The inversion technique characterizes heterogeneities by the departure of the velocity structure beneath the array from a plane-layered reference model. Each depth layer beneath the array is subdivided into blocks and the velocity within each block is allowed to vary with the constraint that the average velocity for a given layer remains fixed. Using the observed station residuals (observed minus calculated arrival times) for many events, at different azimuths and distances and with many intersecting raypaths, one can, in principle, find the relative velocity structure within each layer and delineate local heterogeneities. In a review paper, Aki (1981) notes that such a procedure has been applied to seventeen arrays around the world and has also been used to map structure ranging from local, geothermal sources to the lateral variations beneath whole continents. Thomson and Gubbins (1982) have modified the Aki, et al., method to include amplitude variations. Haddon and Husebye (1978) developed a method based on the joint inversion of *P*-arrival and amplitude data.

All three of the above methods have been applied to the structure beneath NORSAR, and all three found that the depth range with the most lateral heterogeneity was 150-200 km. With its lower viscosity, the asthenosphere is more mobile than the lithosphere, so the asthenosphere is expected to be more homogeneous. It seems reasonable to infer, therefore, that the heterogeneous region is within the continental lithsophere. Hence, having the maximum lateral heterogeneity in the 150-200 km depth range beneath NORSAR is consistent with the Sacks, et al., (1979) interpretation of the velocity reversal at 235 km in the same region being the lithosphere-asthenosphere boundary

These methods all have in common the inability to determinine the *absolute* velocity of a given horizontal layer. That is, only relative lateral variations within the layer over the horizontal region covered by the array are detected. Hence, they cannot be used to determine a LVZ. What they can and have been used for is finding trends of average velocities within a region or between regions. For example, Romanowicz (1979), using the Aki, et al., technique, finds higher average velocities in the depth range 0-250 km in the eastern US compared with the western US, and in a similar study for the upper mantle beneath western Europe (1980), she finds higher velocities in the same depth range beneath the Scandinavian Shield than under the Alps. Hovland, et al., (1981) applied a comparable approach to model the upper mantle beneath central Europe. While these results are consistent with having thicker lithosphere in the eastern US and beneath the shields, they can be used at most only as supporting evidence for such an interpretation.

Sipkin and Jordan (1975, 1976, 1980) suggest that the differences between oceanic regions and continental regions extend to depths in excess of 400 km, based on studies of studies of *S* travel-time anomalies from multiple *ScS* phases, (S waves which reflect one or more times off the core-mantle boundary). Jordan (1975) interprets these results as indicating that the continental root extends to at least 400 km. In contrast, Okal and Anderson (1975), using similar techniques to analyze virtually the same data set, claim that the travel-time anomalies are small enough that the ocean-continent differences extend to only 250 km depth.

Fairhead and Reeves (1977) inferred the thickness of the African lithosphere using a method based on correlations between the Bouguer anomaly and teleseismic *P*-wave travel time residuals (delay times) for African seismograph stations from (presumed) Soviet underground nuclear explosions. They find that stations distant from the East African Rift have delay times which are about 3 seconds less than at stations near the rift system. They also note that the P_n velocities are lower (7.1 to 7.6 km/sec) near the rift while away from the rift they vary from 7.9 to 8.3 km/sec, and they report a spatial correlation with delay times and heat flow. The thicknesses they inferred range from about 50 km beneath the rift to more than 225 km in nort-central Africa. Figure 19 is their tentative lithospheric thickness map for Africa.

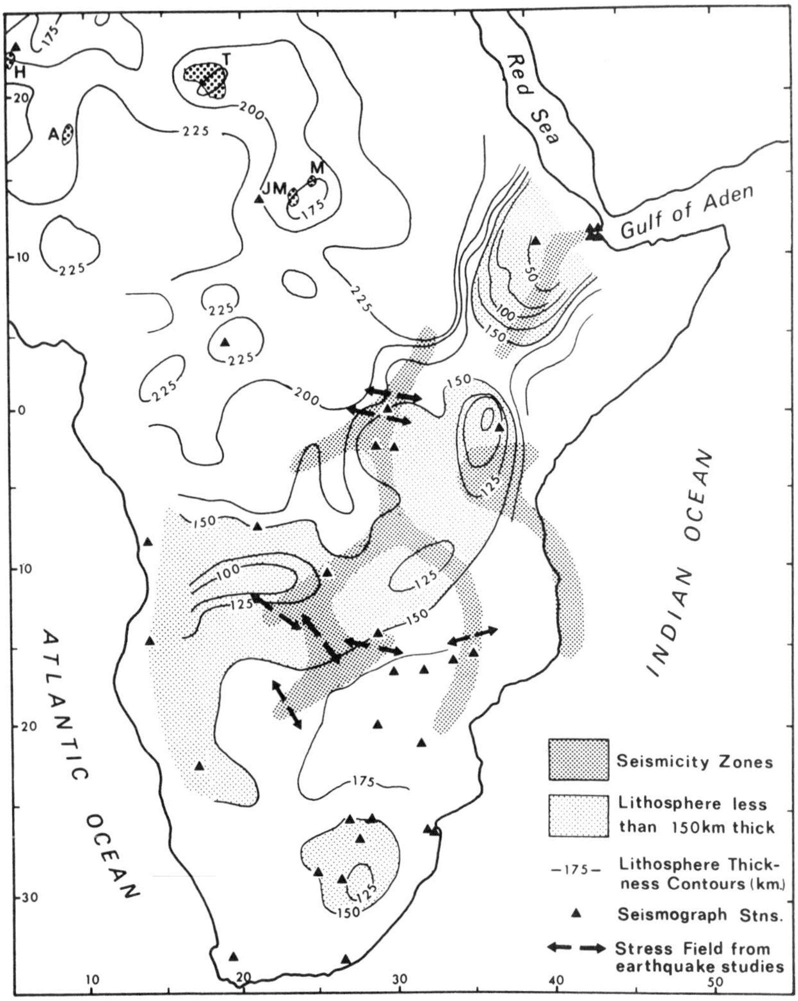

Figure 19: A tentative lithospheric thickness map of Africa based on the interpretation of teleseismic delay times and the regional Bougier anomaly map. Contour intervals are 25 km. Superimposed on the map are zones of Cenezoic volcanism and zones of seismic activity. (After Fairhead and Reeves, 1977.)

Subduction-zone Geometry: In a few cases the *seismicity patterns* themselves provide direct information about the location of a lower plate boundary.

Earthquakes generally occur only within a few tens of kilometers of the upper surface of the lithospheric plate. In subduction regions, one plate underthrusts another, so for at least some distance the upper boundary of the downgoing plate is in contact with the lower boundary or leading edge of another plate. Along some parts of the coast of South America, the subduction geometry, determined principally from seismicity, subducts with a "normal" 30° dip to about 100 km depth. It then becomes horizontal for about 300 kilometers and finally turns downward again at about a 30° dip (Hasegawa and Sacks (1981), our Fig. 20). There is no low Q zone (Sacks and Okada, 1974) and no low velocity zone based on surface-wave studies (James, 1971) in this area at depths less than 200 km, suggesting that the two plates are in contact. Such a pattern occurs beneath both central Peru and central Chile; beneath other regions along the coast the subduction angle is approximately constant at about 30°. A suggested model for such anomalous regions (Sacks 1983) is based on the observation that the density of oceanic lithosphere increases with age. Beneath central Peru, the subducting lithosphere is young enough to be buoyant, assuming that the subducted crustal material has not undergone the density-increasing phase change from basalt to eclogite. The low observed surface heat flow in central Peru suggests that the subducted crustal temperatures may be low enough to retard this phase change - taking into account the kinetics of the process (e.g., Ito and Kennedy. 1971; Ahrens and Schubert, 1975). An implication of that model, then,

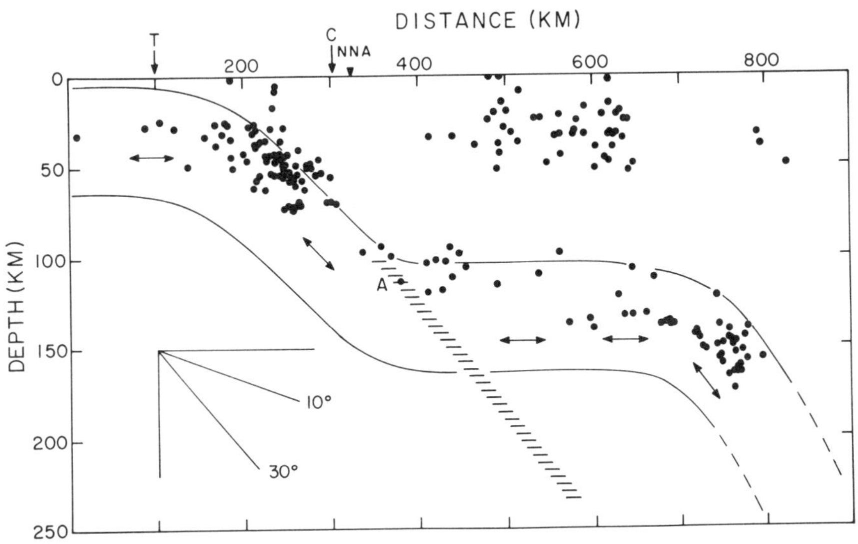

Figure 20: Cross section showing the inferred geometry of the descending Nazca plate beneath central Peru. Seismicity is from Barazangi and Isacks (1979). This is a 1000 km wide section and therefore assumes the deformation of the slab constant over this width. The hatched zone denotes the locues of possible conversion points of the *ScSp* arrivals at station NNA (triangle) estimated by Snoke, et al., (1977). Tension axes, shown by arrows, for focal mechanisms of intermediate-depth earthquakes are approximately parallel to the local dip of the descending Nazca plate. The two plates are assumed to be in contact throughout the horizontal region. The boundary between the plates therefore delineates the base of the continental lithosphere. (After Hasegawa and Sacks, 1981.)

is that in these regions the shape of the subduction zone is dictated by the shape of the lower boundary of the underthrusted South American plate. Accordingly, the continental lithosphere in these regions is interpreted as being about 100 kilometers thick from the coast inland for a horizontal distance of about 300 kilometers. At this point the continental lithosphere thickens, forcing a sharp downturn in the underthrusting plate.

3. TYPES OF CONTINENTAL LITHOSPHERE

Table 1 contains a list of continental regions for which estimates of the lithospheric thickness have been made. In a few cases similar estimates come from independent techniques, but in others the estimates are in controversy; we discuss these in the following paragraphs.

Basins, rift valleys and alpine regions all have reasonably well-developed low-velocity and low-Q zones. The velocity in the LVZ is less than or equal to that in suboceanic asthenosphere at the same depth, i.e., ~4.3 km/sec. at depths greater than 100 km. Some regions, such as the Great Basin in the western US has been studied by various independent techniques - body waves, surface waves, anelasticity - with rather similar results. Yet the structure is rather more ambiguous than the typical oceanic case: based on fundamental-mode Rayleigh wave phase velocity disperesion curves for paths near the East African Rift, which were similar to curves for the Great Basin, Knopoff and Schlue (1972) found they could not distinguish between a structure such as that shown in Fig. 7 and one with a constant velocity of 4.25 km/sec down to a depth of 200 km. Priestley, et al. (1979) established the necessity of the higher lid velocity using higher-mode surface waves. As Priestley, et al., among others, have pointed out, there may well be considerable variation in the crust and upper mantle structure within the Basin and Range province. This heterogeneity might help explain the apparent discrepancy between their model and the model of Biswas and Knopoff (1974), who were unable to get stable solutions with a high lid velocity in their fundamental-mode Rayleigh-wave inversion for paths just to the west of the Rockies.

Aseismic platforms may have a well-developed LVZ, as Biswas and Knopoff (1974) found from a Rayleigh-wave dispersion study of the southern US. They found high lid velocities, as in the northcentral North American shield regions, but a well-developed LVZ starting at a depth between 80 and 140 km with a minimum channel velocity possibly as low as 4.2 km/sec.

Eastern Australia is another region classified as an aseismic platform, but for this region the results are ambiguous. On the one hand, Goncz and Cleary (1976), using a "raw" Monte Carlo inversion of fundamental-mode Rayleigh-wave phase velocities obtained using distant earthquakes, find a structure (depth of 80-120 km to the top of a LVZ) similar to that found for the

Table 1. Depth to Base of Lithosphere

Region	Method	Depth (km)	References
W. United States	P and S wave amplitude	80	Gutenberg (1926, 1948)
W. United States	P wave synthetics	65	Burdick and Helmberger (1978), Burdick (1981)
W. United States	Surface waves	65	Priestley and Brune (1978)
W. United States	Surface waves including higher modes	65	Priestley *et al.* (1980)
Rhine-Alpine	Surface waves	45	Panza *et al.* (1980)
W. Europe	Surface waves	100	Panza *et al.* (1980)
Caledonides	Surface waves	100+	Clark and Stuart (1981)
S. United States	Surface waves	80-140	Biswas and Knopoff (1974)
Japan	Q body waves	70	Sacks (1975)
W. United States	Q body waves	70	Helmberger (1973), Burdick and Helmberger (1978)
W. United States	Q surface waves	80	Lee and Solomon (1975, 1978, 1979)
Eurasia	Q surface waves	125	Burton and Bennell (1976)
East African Rift	Teleseismic P delay, Bouguer anomaly	50	Fairhead and Reeves (1977)
North Central Africa	Teleseismic P delay, Bouguer anomaly	225	Fairhead and Reeves (1977)
W. South America Central Peru, Central Chile	Subduction zone geometry	100	Hasegawa and Sacks (1981)
N.W. and W. Europe	Higher mode surface waves	140	Nolet (1977)
N.W. and W. Europe	Higher mode surface waves	190* density decrease	Nolet (1977)
E. United States	Higher mode surface waves	60	Cara (1979)
E. Australia	Surface waves	80-120**	Goncz and Cleary (1976)

*Cara *et al.* (1980) in a similar analysis do not require the density decrease.

**Mills and Fitch (1977) do not find a low velocity zone in the same region.

southern US. However, Mills and Fitch (1977), using a controlled Monte Carlo technique to invert multimode surface wave data for a moderate earthquake at regional distances found no need for an LVZ to depths of at least 140 km, their maximum resolved depth, as well as relatively low (4.2-4.32 km/sec) subcrustal velocities. These studies are in roughly the same region as Muirhead, et al., (1977) refraction study referred to above, which found a LVZ in *P* in the depth range 155-190 km, somewhat deeper than that found by Goncz and Cleary and below the penetration depth of Mills and Fitch.

The subcrustal velocity beneath Japan has been determined from a number of explosive-source refraction experiments (e.g. Yoshii and Asano, 1972). The P_n velocity is about 7.7 km/sec, i.e. there is no high velocity lid, (The velocity is similar to that of the Basin and Range Province of the western US, Fig. 4.) Shiono, et al., (1980) studied the Japan lithosphere structure using the group velocities of Rayleigh and Love waves. Their preferred structure, fitting the Rayleigh waves over a period range of 10 to 85 seconds is shown in figure 7. A velocity reversal was not consistent with the data. Because the subcrustal velocities are in the range of normal asthenosphere, it might seem plausible to equate the moho and the lithosphere-asthenosphere boundary. However, the anelasticity data show a marked decrease in Q below about 65 km depth, suggesting that that depth is the base of the (somewhat low velocity) lithosphere.

Results for Precambrian shield regions indicate that subcrustal velocities are high, and that if a LVZ exists, it is at most a moderate one starting at a depth greater than 100 km. Table 2 lists regions, including some shields, for which weak LVZ's have been reported.

4. CONCLUDING REMARKS

We are trying to identify the vertical extent of the continental lithosphere, i.e., the continent and its root, which translate as a plate-tectonic unit. Our method for most regions has been to search for asthenosphere which is of the suboceanic type at least so far as velocity and anelasticity are concerned. From the seismological perspective this must be considered an interim report: Such regions as basins, rift valleys, some "young" areas such as western South America, have reliably determined lithospheric thicknesses (see Table 1). For most shields and many aseismic platforms, however, this most basic parameter is uncertain. These regions are characterized by the absence of a pronounced low velocity zone. If a velocity reversal is detected, the minimum shear wave velocity is approximately 4.5 km/sec, well above the 4.3 km/sec found for suboceanic asthenosphere at similar depths. If no "lid" ($V_S \geq 4.6$ km/sec) is required by the analysis, neither is a LVZ. In regions in which there is a reported LVZ with a relatively high velocity and no accompanying

Table 2. Depth to Weak Velocity Reversal

Region	Method	Depth (km)	References
S.E. Australia	P refraction	155	Muirhead et al. (1977)
N. Australia	P refraction	200	Hales et al. (1980b)
E. Siberia	P refraction	200	Yegorkin and Pavlenkova (1981)
W. Australia	Surface waves	200-310, or none	Goncz and Cleary (1976)
Scotland England	Surface waves	140+ or none	Clark and Stuart (1981)
Canadian Shield	Surface waves	115	Brune and Dorman (1963)
N.E. U.S.	Higher mode surface waves	120	Cara (1979)
Baltic Shield	Converted phases	220-250	Sacks et al. (1979)
South American Shield	Converted phases	365-435	Sacks and Snoke (1977)
Baltic Shield	Travel time anomalies	150+ (see text)	Aki et al. (1976) Haddon and Husebye (1978) Thompson and Gubbins (1982)

low Q region (e.g., the Canadian shield), it is not obvious that the LVZ really represents the asthenosphere.

If some continental lithosphere is considerably more than 200 km thick (e.g., Sacks and Snoke, 1977; Jordan, 1975), the detection of the LVZ would be beyond the range of most surface-wave studies. (The depth resolution is improved to some extent if higher modes are used.)

Sufficiently long-line body-wave refraction studies to allow the determination of structure at depths in excess of 200 km have found at most weak LVZ's at such depths (e.g., Yegorkin and Pavlenkova, 1981, beneath the East Siberian platform) or LVZ's which are part of a fairly complicated structure (e.g., Leven, et al., 1981, beneath northern Australia). Such structures would probably not be detectable by surface-wave studies.

There is a potential lack of resolution if the bottom of the lithosphere is at a depth near either the possibily ubiquitous Lehmann discontinuity at about 200 km (e.g., Anderson, 1979) or the 400 km discontinuity. These discontinuities are usually modeled as (fairly gradual) increases of velocity with depth, while the lithosphere is expected to have higher velocity than the deeper material, the velocity changing over at most several kilometers. Converted phases could result from either velocity increases or decreases with depth across an interface, but relative polarity of arrivals can be used to determine the sign of the velocity contrast (e.g., Sacks and Snoke, 1977; Sacks, et al., 1979; Langston, 1979). We note that a rather different velocity model such as the *CAPRI* model shown above in Fig. 2 could explain the

Baltic Shield conversions with a fairly complicated structure in the vicinity of the discontinuity. Further work is required to unravel the fine structure of such interfaces.

It is possible that, unlike the suboceanic situation, subcontinental lithosphere and asthenosphere may not necessarily be of the same material (e.g., Jordan, 1981). There need not be a significant change in the velocity between lithosphere and asthenosphere. Body-wave Q studies may help identify the transition region as was found in Japan. Because a high-velocity lid did not form beneath the Japan moho, surface-wave studies show that there is no significant velocity reversal. Yet a clear anelasticity discontinuity was found at about 65 km depth (Shiono, et al., 1980).

A summary of generalizations concerning the continental lithosphere is as follows:

1) Subcrustal velocities in rift valleys and basins are lower by at least 5% than in shields and most stable platforms.
2) Rift valleys, basins and some aseismic platforms have total lithospheric thicknesses of 45 to 140 km. The sublithospheric velocities are somewhat lower than suboceanic asthenospheric values at the same depth in regions with tensional tectonics (i.e., rift zones and the Great Basin) but beneath aseismic platforms the asthenospheric values are similar or slightly higher than beneath oceans.
3) In well-developed LVZ's, the Q is found to be low ($Q_P \le 200$), as is the case in the suboceanic asthenosphere. At least in one region (Japan), the anelasticity structure determines a lithosphere thickness even though there is no resolvable velocity reversal.
4) Shields have poorly determined thicknesses. There are three possible reasons:
   a: There may be no asthenosphere in the conventional sense. However the continents do translate as a unit.
   b: The base of the lithosphere is too deep to be resolved (to date) and/or it may be indistinguishable from the 200 or 400 km discontinuities.
   c: There may be a very small velocity change at the base of the lithosphere.
5) The results for shield regions and for Japan raise questions concerning the validity of the assumption that the LVZ represents the asthenosphere.

REFERENCES

Ahrens, T. J. and G. Schubert, 1975. Gabbro-eclogite reaction rate and its geophysical significance, *Rev. Geophys. Space Sciences, 13,* 383-440.

Aki, K., 1981. 3-D inhomogeneities in the upper mantle, *Tectonophysics, 75,* 31-40.

Aki, K., A. Cristofferssson, and E. S. Husebye, 1976. Three-dimensional seismic structure of the lithosphere under Montana LASA, *Bull. Seismol. Soc. Amer., 66,* 501-524.

Aki, K., A. Cristofferssson, and E. S. Husebye, 1977. Determination of the three-dimensional seismic structure of the lithosphere, *J. Geophys. Res., 82,* 277-296.

Anderson, D. L., 1967. Phase changes in the upper mantle, *Science, 157,* 1165-1173.

Anderson, D. L., 1979. The deep structure of continents, *J. Geophys. Res., 84,* 7555-7560.

Barazangi, M. and B. Isacks, 1979, Subduction of the Nazca plate beneath Peru: Evidence from spatial distribution of earthquakes, *Geophys. J. R. Astr. Soc., 57,* pp. 537-555.

Baumgardt, D. R., 1981. Seismic body-wave study of vertical and lateral heterogeneity in the Earth's interior, *Ph.D. Thesis,* The Pennsylvania State University, 531 p.

Biswas, N. N. and L. Knopoff, 1974. The structure of the upper mantle under the United States from the dispersion of Rayleigh waves, *Geophys. J. R. Astr. Soc., 36,* 515-539.

Bolt, B. A., 1976. *Nuclear Explosions and Earthquakes: The Partial Veil* (W. H. Freeman, San Fransisco).

Bott, H. P., 1982. *The Interior of the Earth: its structure, constitution and evolution* (Edward Arnold (Publishers) Limited, London).

Brune, J. and J. Dorman, 1963. Seismic waves and earth structure in the Canadian shield, *Bull. Seism. soc. Amer., 53,* 167-210.

Burdick, L. J., 1981, A comparison of the upper mantle structure beneath North America and Europe, *J. Geophys. Res., 86,* 5926-5936.

Burdick, L. J. and D. V. Helmberger, 1978. The upper mantle *P* velocity structure of the western United States, *J. Geophys. Res., 83,* 1699-1712.

Burton, P. W., 1977. Inversions of high frequency $Q_{\gamma}^{-1}(f)$, *Geophys. J. R. Astr. Soc., 48,* 29-51.

Burton, P. W. and J. D. Bennell, 1976. Q^{-1} and lithospheric thickness, *Earth and Planet. Science Letters, 30,* 151-154.

Cara, M., 1979. Lateral variation of *S* velocity in the upper mantle from higher Rayleigh modes, *Geophys. J. R. Astr. Soc., 57,* 649-670.

Cara, M., A. Nercessian and G. Nolet, 1980. New inferences from higher mode data in western Europe and northern Eurasia, *Geophys. J. R. Astr. Soc., 61,* 459-478.

Clark, R. A. and G. W. Stuart, 1981. Upper mantle structure of the British Isles from Rayleigh wave dispersion, *Geophys. J. R. Astr. Soc., 67,* 59-75.

Der, Z., R. Masse, and M. Landisman, 1970. Effects of observational errors on the resolution of surface waves at intermediate distances, *J. Geophys. Res., 75,* 3399-3409.

Fairhead, J. D. and C. V. Reeves, 1977. Teleseismic delay times, Bouguer anomalies and inferred thickness of the African lithosphere, *Earth and Planet. Science Lett., 36,* 63-76.

Given, J. W. and D. L. Helmberger, 1980. Upper mantle structure of northwestern Eurasia, *J. Geophys. Res., 85,* 7183-7194.

Goncz, J. H. and J. R. Cleary, 1976. Variations in the structure of the upper mantle beneath Australia from Rayleigh wave observations, *Geophys. J. R. Astr. Soc.,44,* 507-516.

Gutenberg, B., 1926. Untersuchung zur Frage, bis zu weicher Tiefe die Erde kristallin ist, *Z. Geophys., 2,* 24-20.

Gutenberg, B., 1948. On the layer of relatively low wave velocity at a depth of about 80 kilometers, *Bull. Seismol. Soc. Amer., 38,* 121-148.

Haddon, R. A. W. and E. S. Husebye, 1978. Joint interpretation of *P*-wave time and amplitude anomalies in terms of lithospheric heterogeneities, *Geophys. J. R. Astr. Soc., 55,* 19-44.

Hales, A. L., K. J. Muirhead and J. M. W. Rynn, 1980a. Crust and upper mantle shear velocities from controlled sources, *Geophys. J. R. Astr. Soc., 63,* 659-670.

Hales, A. L., K. J. Muirhead and J. M. W. Rynn, 1980b. A compressional velocity distribution for the upper mantle, Tectonophysics, 63, 309-348.

Hasegawa, A. and I. S. Sacks, 1981. Subduction of the Nazca plate beneath Peru as determined from seismic observations, *J. Geophys. Res., 86,* pp. 4971-4980.

Helmberger, D. V., 1973. On the structure of the low velocity zone, *Geophys. J. R. Astr. Soc., 34,* 251-263.

Hovland, J., D. Gubbins and E. S. Husebye, 1981. Upper mantle heterogeneities beneath central Europe, *Geophys. J. R. Astr. Soc., 66,* 261-284.

Ito, K. and G. C. Kennedy, 1971. An experimental study of the basalt-garnet granulite-eclogite transition, in *The Structure and Physical Properties of the Earth's Crust, Geophys. Monogr. Ser., 14,* edited by J. G. Heacock, pp. 303-314, AGU, Washington, DC.

James, D. E., 1971. Andean crustal and upper mantle structure, *J. Geophys. Res., 76,* 3246-3271.

Jordan, T. H., 1975. The continental tectosphere, *Rev. Geophys. Space Phys., 13,* 1-12.

Jordan, T. H., 1981. Continents as a chemical boundary layer, *Phil. Trans. Roy. Soc. Series A, 301,* 359-373.

Keilis-Borok, V. I. and T. B. Yamovskaja, 1967. Inverse problems of seismology (structural review), *Geophys. J., 13,* 223-234.

King, D. W. and G. Calcagnile, 1976. *P*-wave velocities in the upper mantle beneath Fennoscandia and western Russia, *Geophys. J. R. Astr. Soc.*, *46*, 407-432.

Knopoff, L., 1972. Observation and inversion of surface-wave dispersion, *Tectonophysics*, *13*, 497-519.

Knopoff, L. and F.-S. Chang, 1977. The inversion of surface wave dispersion data with random errors, *J. Geophys.*, *43*, 299-309.

Knopoff, L. and J. W. Schlue, 1972. Rayleigh-wave phase velocities for the path Addis Ababa-Nairobi (in *East African Rifts*, R. W. Girdler (Editor), Elsevier Pub. Co., Amsterdam), *Tectonophysics*, *15*, 157-163.

Langston, C. A., 1977. Corvallis Oregon, crustal and upper mantle receiver structure from teleseismic *P* and *S* waves, *Bull. Seismol. Soc. Amer.*, *67*, 713-724.

Langston, C. A., 1979. Structure under Mount Rainier, Washington, inferred from teleseismic body waves, *J. Geophys. Res.*, *84*, 4749-4762.

Langston, C. A. and D. E. Blum, 1977. The April 29, 1965, Puget Sound earthquake and the crust and upper mantle structure of western Washington, *Bull. Seismol. Soc. Amer*, *67*, 693-711.

Lee, W. B. and S. C. Solomon, 1975. Inversion schemes for surface wave attenuation and Q in the crust and the mantle, *Geophys. J. R. Astr. Soc.*, *43*, *47-71*.

Lee, W. B. and S. C. Solomon, 1978. Simultaneous inversion of surface wave phase velocity and attenuation: Love waves in western North America, *J. Geophys. Res.*, *83*, 3389-3400.

Lee, W. B. and S. C. Solomon, 1979. Simultaneous inversion of surface-wave phase velocity and attenuation: Rayleigh and Love waves over continental and oceanic paths, *Bull. Seism. Soc. Amer.*, *69*, *65-95*.

Leven, J. H., I. Jackson and A. E. Ringwood, 1981. Upper mantle seismic anisotropy and lithospheric decoupling, *Nature*, *289*, 234-239.

Masse, R. P. and S. S. Alexander, 1974. Compressional velocity distribution beneath Scandinavia and western Russia, *Geophys. J. R. Astr. Soc.*, *39*, 587-602.

Mayer-Rosa, D. and St. Mueller, 1973. The gross velocity-depth distribution of *P*- and *S*-waves in the upper mantle of Europe from earthquake observations, *Z. Geophys.*, *39*, 395-410.

Mills, J. M. and T. J. Fitch, 1977. Thrust faulting and crust-upper mantle structure in East Australia, *Geophys. J. R. Astr. Soc.*, *48*, 351-384.

Mitchell, B. J. and R. B. Herrmann, 1979. Shear velocity structure in the eastern United States from the inversion of surface-wave group and phase velocities, *Bull. Seism. Soc. Amer.*,*69*, 1133-1148.

Mohorovicic, A., 1909. Das Beben vom 8.x.1909, *Jb. Met. Obs. Zagreb (Agram)*, *9*, 1-63.

Muirhead, K. J., J. R. Cleary and D. M. Finlayson, 1977. A long-range seismic profile in south-eastern Australia, *Geophys. J. R. Astr. Soc.*, *48*, 509-520.

Nolet, G., 1977. The upper mantle under western Europe inferred from the dispersion of Rayleigh waves, *Z. Geophys.*, *43*, 265-285.

Okal, E. A. and D. L. Anderson, 1975. A study of lateral heterogeneities in the upper mantle by multiple *ScS* travel time residuals, *Geophys. Res. Lett.*, *2*, 313-316.

Panza, G. F., St. Mueller and G. Calcagnile, 1980. The gross features of the lithosphere-asthenosphere system in Europe from seismic surface waves and body waves, *Pageoph*, *118*, 1209-1213.

Patton, H., 1980. Crust and upper mantle structure of the Eurasian continent from the phase velocity and Q of surface waves, *Rev. Geophys. and Space Physics*, *18*, 605-625.

Priestley, K. and J. Brune, 1978. Surface waves and the structure of the Great Basin of Nevada and western Utah, *J. Geophys. Res.*, *83*, 2265-2272.

Priestley, K., J. A. Orcutt, and J. N. Brune, 1980. Higher-mode surface waves and structure of the Great Basin of Nevada and western Utah, *J. Geophys. Res.*, *85*, 7166-7174.

Romanowicz, B. A., 1979. Seismic structure of the upper mantle beneath the United States by three-dimensional inversion of body wave arrival times, *Geophys. J. R. Astr. Soc.*, *57*, 479-506.

Romanowicz, B. A., 1980. A study of large-scale lateral variations of *P* velocity in the upper mantle beneath western Europe, *Geophys. J. R. Astr. Soc.*, *63*, 217-232.

Sacks, I. S., 1975. Anomalous island arc asthenosphere and continental growth, *Carnegie Inst. Wash. Year Book 74*, 256-266.

Sacks, I. S., 1983. The subduction of young lithosphere, *J. Geophys. Res.*, *88*, 3355-3366.

Sacks, I. S. and H. Okada, 1974, A comparison of the anelasticity structure beneath western South America and Japan, *Phys.*Earth*and Planet.*Int., 9, pp. 211-219.

Sacks, I. S. and J. A. Snoke, 1977, The use of converted phases to infer the depth of the lithosphere-asthenosphere boundary beneath South America, *J. Geophys. Res.*, *82*, 2011-2017.

Sacks, I. S., J. A. Snoke and E. S. Husebye, 1979. Lithosphere thickness beneath the Baltic Shield, *Tectonophysics*, *56*, pp. 101-110.

Shiono, K., I. S. Sacks and A. T. Linde, 1980. Preliminary velocity structure of Japanese islands and Philippine Sea from surface wave dispersion, *Carnegie Inst. Wash. Year Book 79*, 498-505.

Sipkin,S. A. and T. H. Jordan, 1976. Lateral heterogeneity of the upper mantle determined from travel times of multiple *ScS*, *J. Geophys. Res.*, *81*, 6307-6320.

Sipkin,S. A. and T. H. Jordan, 1980. Multiple *ScS* travel times in the western Pacific: implications for mantle heterogeneity, *J. Geophys. Res.*, *85*, 853-861.

Snoke, J. A., I. S. Sacks and H. Okada, 1977, Determination of subducting lithosphere boundary by use of converted phases, *Bull. Seism.*Soc.*Am.*, *67*, pp. 1051-1060.

Thomson, C. J. and D. Gubbins, 1982. Three-dimensional lithospheric modeling at NORSAR: linearity of the method and amplitude variations from the anomalies, *Geophys. J. R. Astr. Soc.*, *71*, 1-36.

Tucker, W., E. Herrin and H. W. Freedman, 1968. Some statistical aspects of the estimation of seismic travel times, *Bull. Seism. Soc. Amer.*, *58*, 1243-1260.

Vinnik, L. P., 1977. Detection of waves converted from *P* to *SV* in the mantle, *Phys. Earth and Planet. Inter.*, *15*, 39-45.

Ward, S. N., 1978. Long-period reflected and converted upper-mantle phases, *Bull. Seism. Soc. Amer.*, *68*, 133-153.

Wiggins, R. A., 1972. The general linear inverse problem: implications of surface waves and free oscillations for earth structure, *Rev. Geophys. and Space Phys.*, *10*, 251-285.

Yegorkin, A. V. and N. I. Pavlenkova, 1981. Studies of mantle structure of U.S.S.R. territory on long range seismic profiles, *Phys. Earth and Planet. Inter.*, *25*, 12-26.

Yoshii, T. and S. Asano, 1972. Time-term analyses of explosion seismic data, *J. Phys. Earth*, *20*, 47-57.

MAJOR AND TRACE ELEMENT COMPOSITION OF THE CONTINENTAL LITHOSPHERE

B. L. Weaver and J. Tarney

Department of Geology, University of Leicester, Leicester LE1 7RH, U.K.

1. INTRODUCTION

Estimating the composition of the continental coast is a rather difficult task, perhaps more so than estimating that of the oceanic crust, despite the logistic problems in sampling the latter. This arises partly from the lithological diversity and tectonic complexity of the continental crust, but also from lack of knowledge of its vertical chemical zonation, uncertainty over the nature of the deep crust, whether crustal compositions have changed with time, and a lack of concensus on models for crustal generation which might constrain some of these uncertainties. It is also important to bear in mind that our concepts concerning crustal generation have undergone a fundamental change over the last decade, before which it was fairly widely believed that the continental crust, like the mantle and the core, was somehow a primary feature of the Earth. Isotopic dating, however, has failed to prove the existence of any sialic crust on the Earth before about 3.8 Gyr, and it is clear from the Apollo, Mariner, Viking and Luna missions that the Earth's crust is both fundamentally different from, and considerably younger than, the crusts of other terrestrial planets. It appears to have been generated in a totally different way. This begs the question of whether the Earth ever had a primitive anorthositic crust similar to the lunar highland crust, and if so, how it was destroyed.

Many attempts have been made to estimate the composition of the continental crust and circumvent the problems of the diversity and complexity of its igneous, sedimentary and metamorphic components. Goldschmidt's (1933) estimate was based upon the composition of glacial clays which effectively sampled a large area of the Scandinavian Shield with little intermediate chemical weathering and fractionation. A number of later estimates combined various proportions of basalt and granitic components (Vinogradov, 1962; Taylor, 1964;

Ronov and Yaroshevski, 1969). Eade and Fahrig (1971) and Shaw et al. (1967) compiled averages of many hundreds of samples from the Canadian Shield. Holland and Lambert (1972), however, derived a total crust composition based on a mixture of upper crust (as represented by shield averages) and average Precambrian granulite-facies rocks (those presumed to underlie the major shields) which is significantly different from averages based on simple igneous rock compositions. All these estimates yield a continental crust having an average intermediate composition in terms of major elements with 60-63 percent SiO_2. Table 1 lists some of these average crust compositions.

TABLE 1

Estimates of the composition of the continental crust.

	A	B	C	D	E
SiO_2	61.9	63.1	60.2	61.9	62.5
TiO_2	1.1	0.8	1.0	0.8	0.7
Al_2O_3	16.7	15.2	15.6	15.6	15.6
Fe_2O_3	7.7	6.7	7.9	6.9	6.1
MnO	—	—	0.1	0.1	0.1
MgO	3.5	3.1	3.9	3.1	3.2
CaO	3.4	4.1	5.8	5.7	6.0
Na_2O	2.2	3.4	3.2	3.1	3.4
K_2O	4.2	3.0	2.5	2.9	2.3
P_2O_5	—	—	0.2	0.3	0.2

Average continental crust estimates are from the following:
A - Goldschmidt (1933),
B - Vinogradov (1962),
C - Taylor (1964),
D - Ronov and Yaroshevski (1969),
E - Holland and Lambert (1972).

Further progress from these somewhat objective estimates, particularly as far as the trace element composition of the crust is concerned, can only be made with some constraints provided by models for the petrogenesis of the crust. For instance, the upper crust, discounting the basaltic component, is rather siliceous, and experimental petrologists are reluctant to countenance derivation of such siliceous liquids directly from the mantle by partial melting. Hence there has been a widespread prejudice that the siliceous upper crust has been derived (a) by fractionation from some more basic parent liquid, or (b) by intracrustal melting from a more basic primitive crustal composition. The lower crust then would comprise either cumulates or refractory residues, and would accord with the more basic nature of the lower crust deduced from seismic refraction studies and heat flow data. Geochemical evidence in favor of such models comes from the fact that post-Archaean sediments (and the majority of upper crustal granitic rocks from which they are derived) have rare-earth patterns with prominent negative europium anomalies. This implies that a compensating positive Eu anomaly must be located in the cumulates or refractory residues of the lower continental crust, since melting of mantle with normal mineral assemblages cannot yield liquids with significant Eu anomalies.

An important advance in estimating the average compositions of the upper crust, lower crust and total crust, using such constraints, was made by S.R. Taylor and co-workers (Taylor, 1967, 1977, 1979; Jakes and Taylor, 1974; Taylor and McLennan, 1981). They considered that andesitic island arcs, which are mostly quite young features, represent a volumetrically abundant new addition of primitive continental material from the mantle. Because such arcs are formed at destructive plate margins, they can be regarded as essentially transient features on a geological time scale, and must eventually be accreted laterally during continental collision and orogenesis. Assuming that such arcs have formed regularly throughout geological history, they potentially represent a major mechanism of crustal growth by lateral accretion. If so, then the compositions of island arcs can furnish a reasonable estimate of the bulk continental crust composition. This is the essence of the 'andesite model' for crustal growth; in effect the mean composition of the continental crust can be approximated by the average island arc andesite. Because a wealth of geochemical data now exists for andesites the model provides reasonable estimates of crustal abundances for a wide range of elements. Moreover, the andesite model can be used to estimate the composition of the lower crust if the assumption is made that accreted island arcs suffer intracrustal melting during orogenesis to yield a granodioritic upper crust and a more refractory, residual, lower crust.

The basic features of the andesite model can be illustrated quantitatively using rare-earth elements (Fig. 1 after Taylor and McLennan, 1981). The total crust is represented by the average island arc andesite (Table 2), and the upper crust is represented by the average post-Archaean sediment (Fig. 1). Assuming for simplicity that intracrustal melting yields approximately one-third "upper crust" and two-thirds "lower crust", the rare-earth pattern for the lower crust can be estimated (Fig. 1). It is possible to make estimates for the abundance of other major and trace elements in the lower crust based on the REE data, as has been done by Taylor (1977). Mass balance of major elements yields a rather mafic composition for the lower crust.

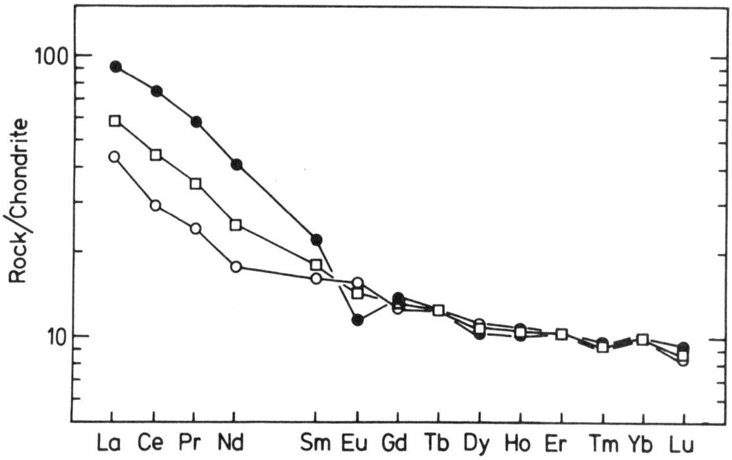

Fig. 1: Abundances of the rare-earth elements (normalized to chondrite abundances of Nakamura, 1974) for the components of the crust in the andesite model. ● - average post-Archaean upper continental crust; □ - average arc andesite; ○ - calculated lower continental crust. Data from Taylor and McLennan (1981).

Because of its apparent success in accounting for the main features of crust compositions, the andesite model has been widely used. Nonetheless, as pointed out by Tarney and Windley (1977), there are certain difficulties: (a) the compositions of studied intraoceanic island arc lavas are significantly more basic than andesite, the "average andesite" being dominated by the more abundant continental margin andesite compositions; thus intraoceanic arcs are too basic to yield the required amounts of typical continental material on remobilization, and (b) there is no positive evidence in any of the major orogenic belts of partial or complete remobilization of accreted island arcs; indeed there are appreciable thermal problems in such remobilization. A more detailed appraisal of the andesite model has

TABLE 2

Estimates of average bulk continental crust composition: A, andesite model (Taylor and McLennan, 1981); B, based on upper crust plus exposed amphibolite- and granulite-facies Archaean crust.

	A	B
SiO_2	58.0	63.7
TiO_2	0.8	0.5
Al_2O_3	18.0	15.8
FeO	7.5	4.7
MnO	0.14	0.07
MgO	3.5	2.7
CaO	7.5	4.5
Na_2O	3.5	4.3
K_2O	1.5	2.0
P_2O_5	—	0.17
Cr	55	61
Ni	30	39
Rb	42	55
Sr	400	498
Ba	350	731
Th	4.8	5.1
U	1.25	1.3
Pb	10	15
Zr	100	213
Nb	11	12
Hf	3.0	4.4
La	19	27
Ce	38	55
Nd	16	23
Sm	3.7	3.9
Eu	1.1	1.07
Tb	0.64	0.50
Tm	0.32	0.23
Yb	2.2	1.46
Lu	0.30	—
Y	22	13
Rb/Sr	0.105	0.110
K/Rb	296	302
Ce_N/Yb_N	4.4	9.7
Eu/Eu*	0.93	0.92

been given by Weaver and Tarney (1982) and further consideration will be given below. In essence, while the model may be geochemically elegant, the associated geological, thermal and tectonic problems are severe.

In seeking a more objective assessment of the composition of the continental crust we need to consider (1) the age of the crust and possible secular variations, (2) geophysical constraints on crustal compositions, (3) evidence from deep crustal xenoliths and exhumed deep crustal terrains bearing on vertical chemical zonation, (4) relevant constraints from experimental petrology, and (5) crust-mantle evolution and compatible crustal growth models.

Several critical geochemical features of the crust also need to be borne in mind: (1) available data suggest that the crust has significant free silica (quartz) - what crustal generation processes can account for this? (2) the upper crust has a marked negative Eu anomaly relative to the mantle - where is the compensating positive Eu anomaly? (3) most crustal compositions have a marked negative Nb-Ta anomaly (with respect to other normally incompatible trace elements) relative to mantle abundances - how has this been achieved?

2. GROWTH OF THE CONTINENTAL CRUST THROUGH TIME AND SECULAR VARIATIONS

In recent years there has been wide acceptance of the view that the continental crust has grown by magmatic additions from the Earth's mantle, at least since 3.8 Gyr ago. A key question is whether the rate of extraction has varied with time, and also whether this has been modified by the recycling of continental material back into the mantle. Isotopic ages yielded by juvenile additions to the crust tend to suggest that crustal growth has been episodic: major periods of crustal generation have been recognized in the time intervals 3.8-3.5, 2.8-2.5, 1.9-1.6, 1.6-1.2, 1.2-0.9, and 0.5-0 Gyr (Moorbath, 1977; Moorbath and Taylor, 1981). Consistent low initial Sr isotope ratios, coupled with the progressive younging of ages away from the older cratonic nuclei, has led to suggestions that crustal growth is irreversible.

On the other hand, based on Wise's (1974) concept of constant freeboard (relative elevation of the continents with respect to sea level) since the Archaean, Armstrong (1968, 1981) has suggested that the continental crust was totally accreted by the end of the Archaean, since when the crust has been continuously and rapidly recycled through the mantle, re-inheriting 'juvenile' mantle isotopic characteristics. Additionally, accelerating continental growth (Hurley and Rand, 1969), quasi-continuous steady growth (Hurley et al., 1962) and even progressive mass loss from the continents via subduction (Fyfe, 1978) have been proposed.

A more realistic view suggests that something over 75 percent of the present volume of continental crust was generated in the Archaean between 3.9 and 2.5 Gyr ago (Dewey and Windley, 1981; McLennan and Taylor, 1982). Since then, net crustal growth has continued, but at a much diminished rate, and with crustal recycling becoming increasingly significant (Dewey and Windley, 1981; Allegre, 1982). For example, the detailed model of Zartman and Doe (1981) for the evolution of Pb isotopes in the crust and mantle demonstrates this feature. In their model ca 70 percent by mass of the continental crust is generated by 2.4 Gyr and net crustal growth since 4.0 Gyr is accompanied by an increasing component of crustal recycling. The mean crustal provenance ages of materials transported to the oceans by rivers, etc., based on Nd isotope model ages (Piepgras and Wasserburg, 1980; Goldstein and O'Nions, 1982) is between 1.0 and 1.5 Gyr, despite the dominant influence of young components in the erosion process, and supports a generally old mean age for the continents. Thus it is important that emphasis is given to Archaean crustal compositions in considering overall models for the continental crust.

A number of workers have demonstrated that there have been significant changes in the compositions of crustal components with time. Engel et al. (1972) noted a change from dominantly sodic magmatic additions in the Archaean to more potassic compositions in the Proterozoic and Phanerozoic. This is matched by an increase in mean initial $^{87}Sr/^{86}Sr$ ratios of granite plutons between the Archaean and Proterozoic which rise well above the mantle Sr isotope growth curve (Glikson, 1979). On the other hand it should be stressed that such isotopic variations correlate broadly with composition and that initial $^{87}Sr/^{86}Sr$ ratios of sodic tonalites and trondhjemites (which dominate the Archaean) follow closely the mantle growth curve irrespective of age (Peterman, 1979). Moreover some of the earliest known crustal rocks (the Amitsoq Gneisses) are quite potassic (Lambert and Holland, 1976).

Nonetheless, secular changes in crustal composition are well documented from the sedimentary record, which averages the chemical diversity of the crust, or more particularly the upper crust. For instance there are notable decreases in heavy metal (Cr, Ni) concentration (Wildeman and Haskin, 1973), increases in rare-earth element abundance (Nance and Taylor, 1976), the development of an increasingly negative Eu anomaly (Jakes and Taylor, 1974), increases in Th and U abundance and Th/U ratio (McLennan and Taylor, 1980), and $^{87}Sr/^{86}Sr$ ratio (Veizer and Compston, 1976) with time. Many of these changes occur quite abruptly at the Archaean/Proterozoic boundary. Their exact significance has still to be elucidated, i.e. whether they reflect fundamental differences in crustal growth processes with time, or a change in balance between competing processes, or simply the fact that the growth of the continental crust during the Archaean established a new geochemical reservoir (including the sub-continental lithosphere) which is decoupled from the convecting mantle reservoir and which produces its own fractionating effects through sedimentary, igneous and metasomatic processes. It is logical to assume, for instance, that the sedimentary component of the continental crust must increase with time unless substantial volumes of sediment have been recycled back into the mantle through subduction processes.

3. GEOPHYSICAL MODELS OF THE CONTINENTAL CRUST

Information derived from seismic reflection and refraction and from heat flow studies provides important constraints on the structure and composition of the crust. Crustal models erected on this basis have been detailed by Smithson and Decker (1974), Smithson and Brown (1977) and Smithson (1978). Recently Fountain and Salisbury (1981) have reviewed the characteristics of exposed crustal sections in the light of geophysical evidence.

Modelling of both seismic and heat flow data suggest an extremely heterogeneous crust, both vertically and laterally (Smithson, 1978), which generally concurs with observations of exposed crustal sections (Fountain and Salisbury, 1981). Even deep crustal sections of similar ages can have different bulk compositions (for instance dominant mafic gneiss in the Archaean Kasila Series; dominant intermediate to silicic gneiss in the equivalent Pikwitonei Belt) with wide variations in the proportions of subsidiary components such as metasediments, anorthosites and ultramafic-mafic bodies. Vertical heterogeneity may in part be due to increasing metamorphic grade with depth. Hence crustal models generally invoke 3 gradational zones: an upper crust of supracrustal sediments invaded by granite, a migmatitic middle crust, and a lower crust of high-grade igneous and metamorphic rocks. Good evidence for the nature of the lower crust is provided by high resolution reflection profiling. For example, in the Wind River Range, Wyoming, the complex deep crustal structure revealed by reflection profiling can be interpreted in terms of exposed Archaean granulites (Smithson et al., 1980).

Seismic velocity data also place constraints on the bulk continental crust composition. The upper crust has seismic velocities appropriate to granodiorite (Smithson, 1978), a composition compatible with most geochemically-based estimates. Although very heterogeneous, average seismic velocities for the lower crust indicate an overall intermediate (diorite) composition, but certainly not a gabbroic composition (Smithson, 1978). However diorite as such need not necessarily be an important component; mixtures of more mafic and more siliceous rock types could equally well be compatible with the seismic data (Smithson, 1978). The mean composition of the bulk continental crust thus approximates to quartz diorite/granodiorite (Smithson, 1978), i.e. is more siliceous than andesite. This is an important observation.

Evidence from heat flow studies shows that the lower crust cannot have similar abundances of radioactive heat producing elements (U, Th, K, Rb) to the upper crust, as observed crustal heat flow is too low (Lachenbruch, 1970; Heier, 1973). Indeed a large proportion of the deep crust must be severely depleted in such elements, particularly in the older shield areas, to explain the heat flow data (Heier, 1973; Smithson and Decker, 1974).

4. DEEP CRUSTAL XENOLITHS

Apart from examining exhumed deep crustal sections preserving high P-T mineral assemblages, xenoliths scavenged from the lower crust by volcanic activity provide evidence of the in-situ petrology and chemistry of the deep crust. A number of such studies have recently been made (Dupuy et al., 1979; Leyreloup et al., 1982; Rogers and Hawkesworth, 1982; Selverstone, 1982; McCulloch et al., 1982).

Best studied are the granulite xenoliths found in the Neogene alkali basalts of the Massif Central (Leyreloup et al., 1977; Dupuy et al., 1979; Dostal et al., 1980), from which it has proved possible to estimate the composition of the lower crust in this region. Metasedimentary material forms a significant proportion (>40%) of the xenolith suite, the remainder being meta-igneous xenoliths which are dominantly mafic in composition and may represent samples of gabbroic intrusions in the lower crust (Dostal et al., 1980). The implied mean lower crustal composition is intermediate (andesitic). Although the heat production is high for lower crustal granulite-facies material (Dostal et al., 1979) it appears that the radioactive heat-producing elements U, K and Rb are somewhat, but not severely, depleted (Fig. 2) relative to the upper crust. The age of these rocks is probably late Precambrian (Dostal et al., 1980).

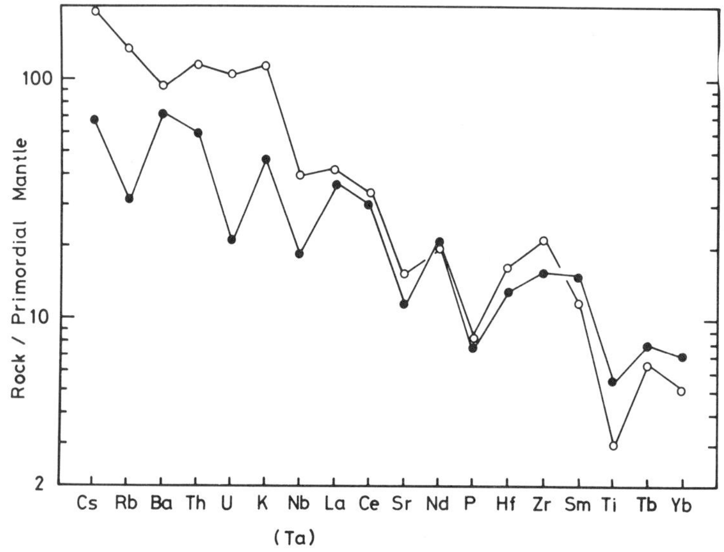

Fig. 2: Abundances of trace elements (normalized to primordial mantle values of Wood et al., 1981) in the average lower continental crust beneath the Massif Central (● - Dupuy et al., 1979) compared to the average upper continental crust (○ - Taylor and McLennan, 1981).

The xenolith suite contained in the Neogene alkali volcanics of Hoggar, Algeria is also a mixture of metasedimentary and meta-igneous granulites (Leyreloup et al., 1982). The metasedimentary xenoliths are all silicic and probably of Archaean age. The meta-igneous xenoliths are quartz-normative calc-alkaline basalts and related plagioclase-rich cumulates, possibly representing a fossil magma chamber of Pan-African or early- to mid-Proterozoic age (Leyreloup et al., 1982).

In southern Africa the granulite-facies xenoliths found in Lesotho kimberlites (Rogers, 1977) are Proterzoic (1.4 Gyr; Rogers and Hawkesworth, 1982) and are almost entirely mafic. Moreover they have the chemical characteristics of cumulates and associated liquids. A similar Proterozoic granulite xenolith suite has been described from kimberlite pipes in South Australia (McCulloch et al., 1982). Again the compositions are mafic, some with the characteristics of cumulates, but a few are more silicic, possibly metasedimentary. The relative abundance of mafic granulites in these and other (mostly Proterozoic) xenolith suites suggests that underplating of the crust by mafic magmas may be an important process, at least during this period of Earth history. Note however that metasedimentary material is an important component too.

Gabbroic granulite xenoliths recovered from alkali basalts in southern Chile have provided information on the composition of the metamorphic fluid (Selverstone, 1982). Fluid inclusion studies indicate the presence of an almost pure CO_2 fluid phase during metamorphism of these lower crustal gabbros. The oldest known crust in this region is 350 $\pm$ 100 Myr (Selverstone, 1982). This is an interesting observation because CO_2 was apparently the dominant fluid phase associated with the production of many older deep crustal granulites, and poses the question of how and why was CO_2 introduced into the deep crust?

5. EXPOSED DEEP CRUSTAL TERRAINS

Although xenolith studies provide general evidence on the in-situ nature of the deep crust, it is not possible to estimate the depth from which they were derived, to know whether they are consanguinous, or to elucidate with confidence petrogenetic relationships. There are few xenolith rock types which do not have counterparts in exhumed high grade terrains. Thus emphasis must be placed on exposed crustal sections where relationships between rock types can be ascertained and petrogenetic models for crustal growth constructed.

An important implication from heat flow studies is that a large proportion of the deeper crust must be depleted in the radioactive heat-producing elements U, Th, K and Rb relative to the upper crust (Heier, 1973). Confirmation of this is provided by the very low concentrations of these elements in many granulite terrains (Tarney and Windley, 1977). Some mechanism is therefore required to redistribute these elements between the lower and upper crust. Two plausible mechanisms have been suggested. Firstly, gross intracrustal partial melting could remove the heat-producing elements (along with other incompatible elements) in a granodioritic liquid from the deep crust, leaving a dry, granulite-facies residuum in the lower crust (e.g. Fyfe, 1973). Secondly, the heat-producing elements might be removed in an active fluid phase during high-grade metamorphism and dehydration of the lower crust, without partial melting taking place (Tarney and Windley, 1977; Newton et al., 1980). A central issue in such processes is the nature and role of the fluid phase. During high-grade metamorphism (P = 8-10Kb; T = 800-900°C for many granulite-facies assemblages) in the presence of a fluid phase with a moderate to high H_2O/CO_2 ratio, partial melting would occur in compositions ranging from granitic to basaltic (Wyllie, 1977). On the other hand the effect of a fluid with low H_2O/CO_2 ratio is to raise solidii temperatures while promoting dehydration reactions (Wyllie, 1977). It is important then to examine the evidence for the nature of the fluid phase in high-grade terrains as well as the chemical evidence for and against partial melting.

Although there is as yet a fairly limited data base, it is becoming increasingly apparent that fluid inclusions in most granulite-facies rocks are CO_2-rich (Touret, 1971, 1974; Selverstone, 1982). Moreover, in regions where there is a variation in metamorphic grade, such as in the Kleivan Granite of southwest Norway, granulite-facies (charnockitic) assemblages have CO_2-rich inclusions whereas inclusions in amphibolite-facies rocks are H_2O-rich (Madsen, 1977; Falkum et al., 1979). These observations, coupled with studies such as those of Janardhan et al. (1979) and Newton et al. (1980), suggest that the development of granulite-facies assemblages is critically dependent on the fluid phase composition.

The chemical effects related to high-grade metamorphism can be best assessed by comparing the composition of high-grade granulite terrains with adjacent, contemporaneous, lower grade amphibolite-facies zones. The effects of partial melting, if important in generating the 'residual' character of granulites should become apparent, particularly from their respective trace element compositions. Weaver and Tarney (1980a, 1981) have recently made such a comparative study of Archaean granulite- and amphibolite-facies gneisses from the Lewisian of northwest Scotland, which represent juxtaposed segments of deep and intermediate crustal levels.

The most striking difference between the two gneiss groups is the much lower concentrations of K, Rb, Th, U and, to a lesser extent, Pb in the granulites, resulting in significantly lower Rb/Sr, K/Sr, Rb/Ba, K/Ba ratios and higher K/Rb ratios in the latter. The granulite-facies average is also more basic, but this is solely a result of the higher proportion of more mafic and ultramafic gneisses in the granulite terrain. At equivalent silica levels in fact granulite- and amphibolite-facies gneisses have similar rare-earth element (REE) and other trace element abundances (with the exception, of course, of K, Rb, Th, and U) and characteristics. The average Lewisian granulite does have a slight positive Eu anomaly (Fig. 3) which could be interpreted as a feature imparted by the removal of a granodioritic melt, leaving a more basic refractory residue. However, it is the more silicic components (tonalites and trondhjemites) of the granulite terrain which carry the positive Eu anomaly, not the mafic gneisses (Fig. 4). Nor is there any significant difference in other trace elements (e.g. Ba) which might be expected to behave incompatibly during melting. It is not possible to explain the differences between the two groups in terms of partial melting models (Weaver and Tarney, 1981). Rather, it is mainly the radioactive heat-producing elements which have been removed from the granulites, probably through the agency of fluid transport. Flushing of CO_2-rich fluids through the lower crust promotes dehydration reactions (cf Tarney et al., 1972) releasing these elements. Other compositional differences between the granulite- and amphibolite-facies zones result mainly from variations in the proportion of mafic and ultramafic to silicic gneisses. This in itself may be controlled by crustal level, with mafic magmas, or tectonically intercalated mafic masses, being restricted to deeper crustal levels by density filtering. It is worth noting that there is still significant free quartz in the silicic gneisses of the granulite terrain, implying that the crust as a whole has appreciable free silica.

Further evidence of the chemical effects associated with granulite facies metamorphism can be gained from the Archaean terrain of southern India. In this terrain metamorphic grade increases to the south, with a well marked amphibolite- to granulite-facies transition in southern Karnataka (Pichamuthu, 1965). Of particular importance is the well-documented occurrence in Kabbaldurga quarry of in-situ transformation

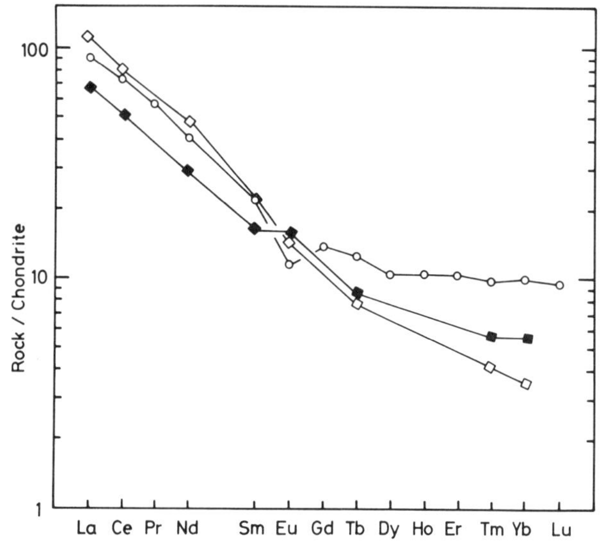

Fig. 3: Abundances of the rare-earth elements (normalized to chondrite abundances of Nakamura, 1974) in the components used to calculate the average composition of the continental crust of Table 2. Average post-Archaean upper crust (O) of Taylor and McLennan (1981); average Lewisian amphibolite facies gneiss (□) of Weaver and Tarney (1981); average Lewisian granulite-facies gneiss (■) of Weaver and Tarney (1980a).

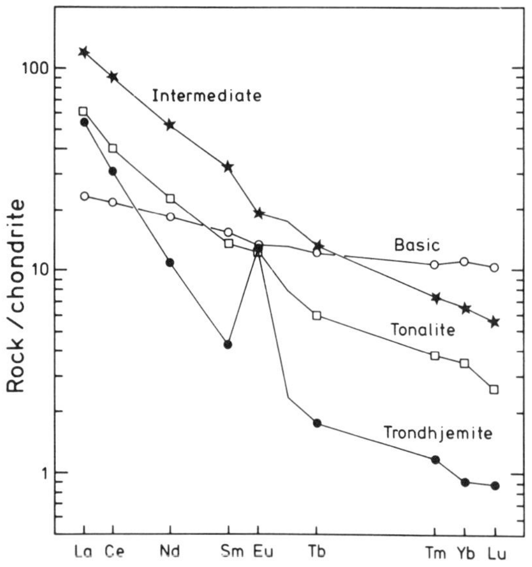

Fig. 4: Abundances of the rare-earth elements (normalized to chondrite abundances of Nakamura, 1974) in the average basic (o), intermediate (★), tonalitic (□), and trondhjemitic (●) gneisses from the Lewisian granulite-facies terrain. From Weaver and Tarney (1980a).

of amphibolite-facies Peninsular gneiss to charnockite (Pichamuthu, 1960; Janardhan et al., 1979; Friend, 1981, 1983). There is no significant difference in either major or trace element chemistry between acid amphibolite- and granulite-facies (charnockitic) gneisses at Kabbaldurga (Weaver and Tarney, 1983). In marked contrast to Scottish Lewisian gneisses, both amphibolite- and granulite-facies silicic gneisses appear to be somewhat enriched in radioactive heat-producing elements over the abundances of other trace elements. On the other hand, mafic gneisses at Kabbaldurga seem to preserve evidence of both trace element enrichment and depletion events. The elements particularly affected are the radioactive heat-producing elements, plus additionally Ba and Sr (Weaver and Tarney, 1983).

These observations are compatible with the model proposed by Weaver (1980) to explain the geochemistry of Madras granulites, which represent a deeper crustal level in the southern Indian shield where granulite-facies metamorphism was pervasive. Weaver (1980) postulated that granulite-facies metamorphism was preceded in southern India by metasomatism by a hydrous fluid phase, with accompanying partial melting and migmatization. However, a change to a CO_2-rich fluid composition may have caused a rapid inversion to granulite-facies assemblages. The concept of flushing of the lower crust by mantle derived CO_2-rich fluids would be an adequate mechanism to explain these features. As the CO_2-rich fluid front ascends through the crust, the leading edge would become enriched in radioactive heat producing elements, as well as H_2O, removed from the deeper levels of the crust during granulite-facies metamorphism. Thus, at some structural level the leading edge of the front might have the composition of a metasomatizing hydrous fluid. Lowering of the solidii by the influx of such a fluid would cause partial melting with accompanying metasomatism. However, continued movement of the ascending front would cause a rapid change in fluid composition to CO_2-rich. This would result in dehydration and crystallization of the partial melts and restites under granulite-facies conditions, possibly without significant melt migration from the zone of partial melting.

Such a mechanism might be capable of explaining the apparently disparate geochemistry of both the Lewisian terrain and the Archaean of southern India. In the Lewisian, juxtaposition of the granulite-facies block against the higher level amphibolite-facies blocks may have removed evidence for the occurrence of a metasomatic migmatite zone. Hence the granulite-facies gneisses represent a deep crustal level which has been purged of radioactive heat-producing elements by

CO_2-rich fluids and the amphibolite-facies represents a much higher crustal level which is largely unmodified by metamorphic processes. In marked contrast, the granulite-facies terrain of the southern Indian shield shows clear evidence for metasomatism and partial melting, with the development of an extensive migmatite zone. A crucial difference between the terrains may be the proportion of metasediment present. In the Lewisian metasedimentary material comprises a trivial component of a terrain dominated by mafic and tonalitic orthogneisses. However, in southern India metasediments (khondalites) form a significant proportion of the gneisses. Such sedimentary material was presumably originally moderately hydrous, and during high-grade metamorphism and flushing by CO_2-rich fluids dehydration would buffer the fluid composition to a higher H_2O/CO_2 ratio. This might well allow extensive migmatization to occur.

The above arguments suggest that gross intracrustal partial melting need not be invoked to explain the large scale chemical differentiation of the crust, and indeed such a process is not supported by the chemical data. Rather, the composition of metamorphic fluids would appear to control the distribution of radioactive heat-producing elements in the crust, while variation in the proportions of rock types may control the major and trace element chemistry as well as modifying effects due to variable fluid composition.

Not all Archaean granulites are depleted in potassium and other heat-producing elements, as the Madras example demonstrates. Large tracts of potassic charnockitic gneisses make up the Brazilian Shield in Bahia State and have the geochemical characteristics of upper crustal granites, having negative Eu anomalies, high Rb/Sr ratios and low K/Rb ratios (Sighinolfi et al., 1981). The simplest explanation for such granulites is that CO_2-rich fluids have risen high in the crust and have promoted the dehydration reactions leading to granulite-facies assemblages. The lack of significant depletion in K and Rb can be linked to the fact that few of the gneisses are diopside-normative, a necessary condition for reactions leading the loss of these elements (Tarney et al., 1972).

THE SEDIMENTARY COMPONENT OF THE CONTINENTAL CRUST

Sedimentary rocks dominate the present surface of the continents. However, it is interesting to note that sedimentary rocks consistently form a small but significant proportion of most exposed Archaean deep crustal granulite terrains, intermixed with gneisses of plutonic origin with mineral assemblages which record P-T conditions of the order of 10Kb and 900°C (Tarney and Windley, 1977). Metasedimentary rocks appear to form an even higher proportion of the crustal xenoliths of Proterozoic age described above. In the early Phanerozoic granulites which are locally uplifted and exposed throughout much of central Europe, metasedimentary rocks are dominant and are interlayered with eclogites and garnet peridotites. We may reasonably ask by what mechanism sedimentary rocks, apparently mostly quartzites and pelites with occasional marbles of shallow water origin, are scattered throughout the continental crust in this way? The proportion of metasedimentary relative to meta-igneous rocks in deep crustal sections appears, as a first approximation, to increase from the Archaean to the present.

The explanation may lie in the processes of sedimentary accretion at active continental margins. It is now well known that in some arc systems such as the Marianas, ocean-floor sediments are not scraped off in the trench but must be subducted into the mantle or underplated beneath the arc. However, where the sediment supply is greater such that the plate is depressed approaching the trench, sediments are scraped off to form an accretionary wedge. This can reach major proportions, with sediments forming the full crustal thickness, perhaps for several hundred kilometers in the case of the late Paleozoic-early Mesozoic accretionary wedge in southern Chile (Forsythe, 1982). Calc-alkaline magmatic products may of course be emplaced into this metasedimentary basement, resulting in a complex mixture of magmatic and sedimentary crustal components at both deep and shallow crustal levels. This sedimentary component of the continental crust might be expected to increase with time in proportion to the increasing area of continent, but there is no simple way of estimating its magnitude.

6. AN ESTIMATE OF THE COMPOSITION OF THE CONTINENTAL CRUST

As we have seen, many workers have attempted to estimate the bulk composition of the continental crust, approaching the problem from several directions. The major uncertainty is the nature of the lower crust. The most recent and comprehensive estimate, including trace element data, is that of Taylor and McLennan (1981), and is based on the average arc andesite (Table 2, column A). Knowledge of the bulk composition is vital in order to assess the degree of chemical mass transfer within the crust, and also the net transfer between mantle and crust with implications for isotope systematics in global chemical geodynamics (DePaolo, 1979; Weaver and Tarney, 1980b).

We believe that quantification of the composition of the crust must be approached empirically, taking account of all available geophysical and chemical constraints and the age distribution of the crust. Particularly important is the implied overall intermediate composition and low radioactive heat production for the lower crust. Equally important is the implied rather siliceous (quartz-diorite to granodiorite) composition for the total continental crust.

Laying emphasis on Archaean crustal compositions, we take the relative proportions of upper, middle and lower crust to be 2:1:3; i.e. in a 36 km thick continental segment the respective thicknesses would be 12, 6 and 18 km. The composition of the upper crust is fairly well known through studies of sediments and from detailed sampling of shield areas, and we take the composition proposed by Taylor and McLennan (1981) as representative of this segment of crust. For the lower crust we note the coherence of observed petrological and chemical features of Archaean granulites with geophysical requirements for the lower crust. The 2.9 Gyr Lewisian granulite complex of northwest Scotland has the requisite major element composition and heat flow, and is extremely well characterized chemically (Sheraton et al., 1973; Holland and Lambert, 1973; Pride and Muecke, 1980; Weaver and Tarney, 1980a). We therefore take the Lewisian granulite average of Weaver and Tarney (1980a) as being an appropriate lower crust composition. An average composition for the migmatitic middle crust is more difficult to assess. However, we take the Lewisian amphibolite-facies average of Weaver and Tarney (1981) because these gneisses appear to represent an intermediate crustal level. They are not, though, true migmatites and the abundances of U, Th, K and Rb, and hence heat production, are rather less than suggested for the migmatitic middle crust (Smithson and Decker, 1974). Rare earth element patterns for these three components are shown in Fig. 3. Combining these data sets, with appropriate weighting, yields the crustal average presented in Table 2, column B. The bulk crust average (average andesite) of

Taylor and McLennan (1981) is included for comparison in column A.

The bulk crustal composition derived through adopting a Lewisian granulite lower crustal component is significantly more siliceous than that of Taylor and McLennan (1981), and has a major element composition compatible with geophysical evidence. This results from considering the lower crust as having an average intermediate rather than the residual mafic composition supposed by the andesite model. The abundances of a number of trace elements, notably Cr, Ni, Rb, Sr, Th, U and Nb, and ratios such as Rb/Sr and K/Rb are comparable in the two models (Table 2). However, the calculated concentrations of other trace elements, particularly the REE, Ba and Zr are appreciably different. The REE pattern is more fractionated (Fig. 5, Table 2), with light REE being higher and heavy REE being lower than in the andesite model. In neither model is there a significant Eu anomaly, allowing for the inflection of the REE patterns.

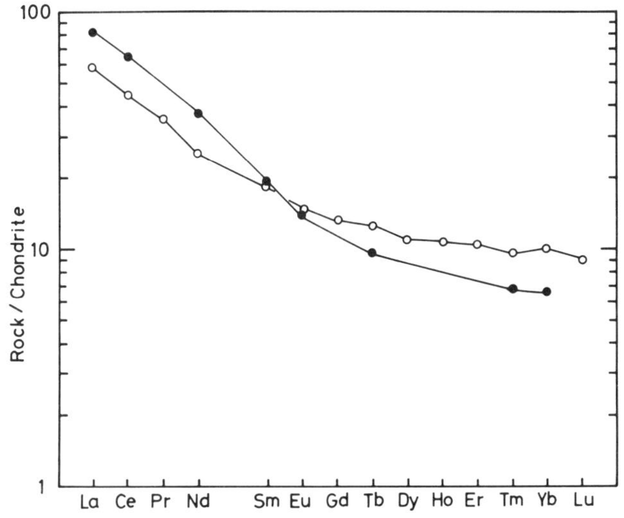

Fig. 5: Abundances of the rare-earth elements (normalized to chondrite abundances of Nakamura, 1974) in the crustal averages of Table 2. ○ - average continental crust of Taylor and McLennan (1981); ● - average continental crust of this paper.

If the continental crust is regarded as the complementary reservoir to the depleted ('MORB') mantle reservoir, the mean Rb/Sr and Nd/Sm (cf Ce_n/Yb_n) ratios in the crust must balance those of the latter reservoir in terms of Sr and Nd isotope characteristics. As pointed out by Weaver and Tarney (1980b), there are significant problems with an 'andesitic' model crust in this regard; however the crustal average reported here, incorporating a depleted granulite lower crust, demonstrably provides a much better fit.

The geochemical characteristics of continental crust can be illustrated (Fig. 6) by plotting the two crustal averages in Table 2 normalized to primordial mantle values (cf Wood et al., 1979). The elements are arranged in order of increasing incompatibility (relative to typical mantle assemblages) from right to left. The differences between the two crustal averages are apparent; but both display some characteristic features. There is a progressive increase in the relative abundance of the more incompatible elements. However, there are significant negative anomalies in the abundances of the high field strength incompatible elements such as Nb (Ta), Ti and P relative to the REE. Such features are characteristic of Phanerozoic granitoids as well as the earliest Archaean granitoid crustal components (Weaver and Tarney, 1980a), and are indeed also characteristic of most modern subduction-related magmas (Saunders et al., 1980).

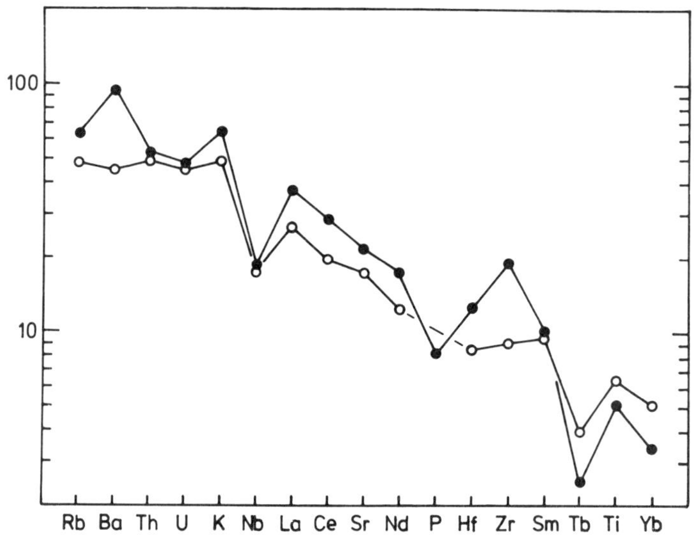

Fig. 6: Abundances of trace elements (normalized to primordial mantle values of Wood et al., 1981) in the crustal averages of Table 2. ○ - average continental crust of Taylor and McLennan (1981); ● - average continental crust of this paper.

7. DISCUSSION

The geophysical and geochemical characteristics of the continental crust described above provide some constraints on crustal generation models. Critical features include the siliceous nature of the bulk crust, its vertical zonation with respect to radioactive heat-producing elements and mafic components, the consistent 'negative' anomalies of Ta and Nb, the 'negative' Eu anomaly in the upper crust, and the increasing proportion of the metasedimentary component of the crust with time.

The pronounced negative anomalies in the abundances of the high field strength elements Nb and Ta (and to a lesser extent Ti) relative to other incompatible elements appears to be a characteristic of all continental crust from the Archaean to the present. It is not a feature of ocean floor or ocean island basalts, or of the primitive lunar crust, but is a consistent feature of all subduction-related magmas. The simplest explanation is that these elements are retained in the source regions of subduction-zone magmas as a result of the enhanced stability of a Ti-rich phase such as rutile, ilmenite or sphene (Saunders et al., 1980) under hydrous conditions. The implication is that water has played an important role in the generation of the continental crust, and moreover that the Earth's continental crust has been generated in a totally different way from the primitive crusts of the other terrestrial planets. Any such primitive crust that the Earth may have had has not been preserved.

Melting of the mantle under conditions of low pH_2O invariably yields basalt, and whereas basaltic magmas can differentiate by crystal fractionation to yield small amounts of silicic liquids, they also produce considerably more crystal cumulates. Such large volumes of cumulates have never been located in exposed deep crustal sections or xenolith suites. Apart then from proposing that such crystal fractionation occurs entirely within the mantle itself, with only the more siliceous differentiates rising into the crust, it is necessary to consider the generation of more siliceous primary magmas from the mantle system.

Wyllie (1982) has recently summarized the experimental evidence relating to the production of more siliceous magmas in subduction zones. He concludes that wet melting of mantle peridotite can only yield magmas as siliceous as andesite at depths shallower than 40 km and where temperatures exceed 1000°C. Such conditions are unrealistic in modern subduction zones. Indeed, thermal models of modern subduction zones (Anderson et al., 1978) indicate that melting of the subducted ocean crust is unlikely owing to the endothermic effects accompanying dehydration of the slab and low geothermal gradients.

Higher geothermal gradients in the past, particularly in the Archaean, may however have permitted hydrous melting of the slab and the liquids generated by hydrous melting of amphibolite, garnet-amphibolite or eclogite can be more siliceous than andesite (Wyllie, 1982). An important feature of subduction zones is the element transfer that accompanies dehydration. The lithophile elements K, Rb, Ba, etc. are the most mobile, and transfer of these elements from the slab to the overriding mantle wedge enables phlogopite and/or hornblende to develop as a stable mantle mineral (cf Wyllie and Sekine, 1982). Moreover, Mysen (1979) has also argued that light rare-earth elements can be transferred from subducted eclogite to the mantle wedge by hydrous fluids. Equally important is that silica is very soluble in superheated fluids, and as subducted ocean basalts would recrystallize, at depths exceeding 80 km, to eclogite bearing appreciable modal quartz, it is likely that much of this free silica would be transferred to the mantle wedge and could significantly modify its composition. In particular, this may permit the generation of more siliceous primary magmas than is otherwise possible by melting of either subducted ocean crust or peridotite mantle. Explaining the rather siliceous nature of observed lower and upper crust compositions (Table 2) would be extremely difficult using basaltic or andesitic magmas as parental liquids. Clearly, hydrous fluids may have played a major role in determining two important features of continental crust, (a) the enhanced concentrations of lithophile elements relative to high field strength elements and (b) the relatively high silica content of the bulk crust.

Petrogenetic models for Archaean high-grade granulite- and amphibolite-facies gneiss terrains have been considered by Weaver and Tarney (1980, 1981). They argued that the dominant tonalitic gneisses had been generated from a mafic source by high pressure fractionation processes involving residual hornblende and/or garnet under high pH_2O. Conversely, the compositions of the associated mafic gneisses could be related to low-pressure fractional crystallization under conditions of low pH_2O. Moreover, the other components of high-grade gneiss terrains (such as the layered anorthosite complexes and ultramafic complexes) also appear to have evolved through low-pressure processes (Weaver et al., 1981, 1982), a deduction confirmed by their common association with metasediments. The tectonic model most compatible with these results is that of an Andean-type cordilleran belt (Tarney, 1976). Tonalitic-trondhjemitic magmas are generated by hydrous partial melting of subducted oceanic crust, this being permitted by the higher geothermal gradients in the Archaean, and these magmas are dominantly added to the crust by underplating. The lenses of mafic gneiss, anorthositic and mafic-ultramafic complexes - the 'low-pressure' components - represent fragments of oceanic lithosphere which have been tectonically emplaced by lateral accretion into the magmatic continental margin. The metasediments may also represent ocean floor or continental margin sediments which have been tectonically accreted in this way. Similar metasediments (psammites, pelites and marbles) occur within the deep level foliated tonalites and diorites of the Patagonian batholith and form part of a pre-existing accretionary wedge. An important feature of such a model is that the mafic-ultramafic components are dominantly introduced into the deeper levels of the accreting continental margin, thus providing a ready explanation for increasing crustal density with depth.

The make-up of newly formed continental crust may have changed with time. The decreasing geothermal gradient since the Archaean would have the net result of decreasing the magmatic component arising from direct melting of subducted ocean crust. At the same time the increasing continental volume generates an increasing volume of upper crust sediment which is ultimately accreted at continental margins. For these reasons the sedimentary component of the crust may have increased with time relative to the magmatic component, a feature which is evident in comparisons of Archaean and Phanerozoic crustal sections and xenolith suites. Nonetheless, if subduction remains the dominant mechanism for both magmatic and sedimentary accretion, tectonic emplacement of mafic material into the deep crust, as well as mafic magma addition, are also viable mechanisms for increasing the density of the lower crust.

The progressive development of a negative Eu anomaly in post-Archaean sediments, reflecting the composition of the upper crust, has been taken to indicate an increasing degree of intracrustal melting since the Archaean, with the compensating positive Eu anomaly being located in the lower continental crust (Taylor and McLennan, 1981). Both the bulk crustal averages of Taylor and McLennan (1981) and that presented here (weighted toward Archaean compositions) lack significant Eu anomalies. Although deep crustal Archaean granulites have a net positive Eu anomaly (Pride and Muecke, 1980), which might in general terms support intracrustal melting, we have stressed (Weaver and Tarney, 1980, 1982) that it is the siliceous tonalitic-trondhjemitic gneisses (having the major element compositions appropriate to partial melts) which impart the positive Eu anomaly. Their petrogenesis is more easily explained in terms of partial melting of a mafic source under conditions of high pH_2O, the positive Eu anomaly being a consequence of residual hornblende or garnet in the source (Weaver and Tarney, 1980); thus the positive Eu anomaly is a primary rather than a secondary feature of the Archaean crust. The positive Eu anomalies in Proterozoic mafic granulite xenoliths in Lesotho kimberlites (Rogers, 1977) have now been reinterpreted as plagioclase-rich cumulates (Rogers and Hawkesworth, 1982). It would seem, therefore, that alternative mechanisms are required to explain the apparent depth dependency of Eu anomalies in the crust.

An important feature of hydrous melts is that they intersect their solidii at higher pressures than dry melts, and therefore crystallize and solidify deep in the crust (Harris et al., 1970). The primary phase volume of plagioclase contracts significantly at high water pressure, whereas that of hornblende expands (Wyllie, 1982). Hence tonalitic-trondhjemitic liquids which have been generated at high pH_2O in equilibrium with residual hornblende, or which have suffered hornblende crystallization, will develop positive Eu anomalies and highly fractionated REE patterns with significant HREE depletion in response to hornblende fractionation (Arth et al., 1978). Such melts cannot, however, easily evolve beyond the deep crust and, unless significant underplating occurs, the plutonic rocks formed from them, and which carry appreciable positive Eu anomalies, must reside in the deeper crust. Conversely, equivalent granitoid magmas generated under low pH_2O conditions can rise to high crustal levels. They will have no Eu anomaly or, if they are generated in equilibrium with plagioclase or they suffer plagioclase fractionation during their long ascent to the surface, they will develop negative Eu anomalies. It is the rocks formed from these high level magmas which dominantly contribute to the sedimentary budget, and which produce the 'negative Eu' signature of the post-Archaean sedimentary record. This signature will be enhanced if sediments which are accreted laterally into the crust by subduction processes are remobilized during later magmatism: as we have seen, the sedimentary component of the crust appears to increase with time relative to the magmatic component.

The significant point arising from the above model, compared with that developed by Taylor and McLennan (1981), is that the upper crustal negative Eu anomaly and the positive Eu anomaly of the lower crust are not directly complementary, but have developed independently as a consequence of whether crustal magma generation occurred under high or low pH_2O. It is not necessary to invoke gross intracrustal melting. But the model emphasizes once again the important role of fluids in determining the geochemical characteristics of continental crust.

In summary, the Earth's continental crust appears to have been formed in a fundamentally different way from the primitive sialic crusts of other terrestrial planets. Several critical geochemical features point to the important influence of hydrous fluids during crustal generation, and by implication, that subduction has been the dominant mechanism through which crustal growth takes place. The important question remains as to why there was apparently no crustal generation before ca 3800 Myr. One can only speculate that (a) previous to this date the planet had not cooled sufficiently since its initial formation for the hydrosphere to form or (b) that the hydrosphere is a secondary feature, perhaps related to the major Imbrium impact features seen on the Moon and other terrestrial planets. Whatever the reason, water would have had a tremendous catalytic effect on the rate of production of continental sial.

REFERENCES

Allègre, C. J. (1982) Chemical geodynamics. Tectonophysics, 81, 109-132.

Anderson, R. N., Delong, S. E. and Schwarz, W. H. (1978) Thermal model for subduction with dehydration in the downgoing slab. J. Geol., 86, 731-739.

Armstrong, R. L. (1968) A model for the evolution of strontium and lead isotopes in a dynamic Earth. Rev. Geophys., 6, 175-199.

Armstrong, R. L. (1981) Radiogenic isotopes: evidence for continental growth or recycling? The case for a near-steady-state no-continental growth crustal-recycling Earth. Phil. Trans. R. Soc., A301, 443-472.

Arth, J. G., Barker, G., Peterman, Z. E. and Friedman, I. (1978) Geochemistry of the gabbro-diorite-tonalite-trondhjemite suite of southwest Finland and its implications for the origin of tonalitic and trondhjemitic magmas. J. Petrol., 19, 289-316.

De Paolo, D. J. (1979) Implications of correlated Nd and Sr isotopic variations for the chemical evolution of the crust and mantle. Earth Planet. Sci. Lett., 43, 201-211.
Dewey, J. F. and Windley, B. F. (1981) Growth and differentiation of the continental crust. Phil. Trans. R. Soc., A301, 189-206.
Dostal, J., Dupuy, C. and Leyreloup, A. (1980) Geochemistry and petrology of meta-igneous granulitic xenoliths in Neogene volcanic rocks of the Massif Central, France - implications for the lower crust. Earth Planet. Sci. Lett., 50, 31-40.
Dupuy, C., Leyreloup, A. and Vernieres, J. (1979) The lower continental crust of the Massif Central (Bornac, France) - with special references to REE, U and Th composition, evolution, heat-flow production. Phys. Chem. Earth, 11, 401-415.
Eade, K. E. and Fahrig, W. F. (1971) Geochemical evolutionary trends of continental plates - a preliminary study of the Canadian Shield. Geol. Surv. Can. Bull., 179.
Engel, A. E. J., Itson, S. P.,, Engel, C. G., Stickney, D. M. and Cray, E. J. (1974) Crustal evolution and global tectonics: a petrogenetic view. Bull. Geol. Soc. Am., 85, 843-858.
Falkum, T., Wilson, J. R., Petersen, J. S. and Zimmerman, H. D. (1979) The intrusive granites of the Farsund area, South Norway: their interrelations and relations with the Precambrian metamorphic envelope. Norsk. Geol. Tidsskr., 59, 125-139.
Forsythe, R. (1982) The late Palaeozoic to early Mesozoic evolution of southern South America: a plate tectonic interpretation. J. Geol. Soc. Lond., 139, 671-682.
Fountain, D. M. and Salisbury, M. H. (1981) Exposed cross-section through the continental crust: implications for crustal structure, petrology and evolution. Earth Planet. Sci. Lett., 56, 263-277.
Friend, C. R. L. (1981) Charnockite and granite formation and influx of CO_2 at Kabbaldurga. Nature, 294, 550-552.
Friend, C. R. L. (1983) The relationship between charnockite formation, granite production and influx of CO_2: evidence from Kabbaldurga, Karnataka, South India. In: Atherton, M. P. and Gribble, C. D. (eds.), Migmatites, melting and metamorphism. Shiva Publ., Nantwich, 264-276.
Fyfe, W. S. (1973) The granulite facies, partial melting and the Archaean crust. Phil. Trans. R. Soc., A273, 457-462.
Fyfe, W. S. (1978) The evolution of the Earth's crust: modern plate tectonics to ancient hot spot tectonics? Chem. Geol., 23, 89-114.
Glikson, A. Y. (1979) Early precambrian tonalite-trondhjemite sialic nuclei. Earth Sci. Rev., 15, 1-73.
Goldschmidt, V. M. (1933) Grundlagen der quantitativen geochemie. Fortsch. Min. Krist. Petrog., 17, 112.
Goldstein, S. and O'Nions, R. K. (1982) Nd isotopic study of river particulates, atmospheric dusts and pelagic sediments. EOS, 63, 352-353.
Harris, P. G., Kennedy, W. Q. and Scarfe, C. M. (1970) Volcanism versus plutonism - the effect of chemical composition. In: Newall, G. and Rast, N. (eds.), Mechanism of igneous intrusion. Geol. J. Spec. Issue No. 2, 187-200.
Heier, K. S. (1973) Geochemistry of granulite facies rocks and problems of their origin. Phil. Trans. R. Soc., A273, 429-442.
Holland, J. G. and Lambert, R. St. J. (1972) Major element chemical composition of shields and the continental crust. Geochim. Cosmochim. Acta, 36, 673-683.

Holland, J. G. and Lambert, R. St. J. (1973) Comparative major element geochemistry of the Lewisian of the mainland of Scotland. In: Park, R. G. and Tarney, J. (eds.), The early Precambrian of Scotland and related rocks of Greenland. University of Keele, Keele, 51-62.
Hurley, P. M., Hughes, H., Faure, G., Fairbairn, H. W. and Pinson, W. H. (1962) Radiogenic strontium-87 model of continent formation. J. Geophys. Res., 67, 5315-5333.
Hurley, P. M. and Rand, J. R. (1969) Pre-drift continental nuclei. Science, 164, 1229-1242.
Jakes, P. and Taylor, S. R. (1974) Excess europium content in Precambrian sedimentary rocks and continental evolution. Geochim. Cosmochin. Acta, 38, 739-745.
Janardhan, A. S., Newton, R. C. and Smith, J. V. (1979) Ancient crustal metamorphism at low pH_2O: charnockite formation at Kabbaldurga, South India. Nature, 278, 511-514.
Lachenbruch, A. H. (1970) Crustal temperature and heat production: implications of the linear heat-flow relations. J. Geophys. Res., 75, 3292-3300.
Lambert, R. St. J. and Holland, J. G. (1976) Amitsoq gneiss geochemistry: preliminary observations. In: Windley, B. F. (ed.), The early history of the Earth. Wiley, New York, 191-201.
Leyreloup, A., Bodinier, J. L., Dupuy, C. and Dostal, J. (1982) Petrology and geochemistry of granulite xenoliths from Central Hoggar (Algeria) - implications for the lower crust. Contrib. Mineral. Petrol., 79, 68-75.
Leyreloup, A., Dupuy, C. and Andriambololona, R. (1977) Catazonal xenoliths in French Neogene volcanic rocks: constitution of the lower crust. Contrib. Mineral. Petrol., 62, 283-300.
Madsen, J. K. (1977) Fluid inclusions in the Kleiven granite, South Norway, I: Microthermometry. Am J. Sci., 277, 673-696.
McCulloch, M. T., Arculus, R. J., Chappell, B. W. and Ferguson, J. (1982) Isotopic and geochemical studies of nodules in kimberlite have implications for the lower continental crust. Nature, 300, 166-169.
McLennan, S. M. and Taylor, S. R. (1980) Th and U in sedimentary rocks: crustal evolution and sedimentary recycling. Nature, 285, 621-624.
McLennan, S. M. and Taylor, S. R. (1982) Geochemical constraints on the growth of the continental crust. J. Geol., 90, 347-361.
Moorbath, S. (1977) Ages, isotopes and evolution of Precambrian continental crust. Chem. Geol., 20, 151-187.
Moorbath, S. and Taylor, P. N. (1981) Isotopic evidence for continental growth in the Precambrian. In: Kroner, A. (ed.), Precambrian Plate Tectonics. Elsevier, Amsterdam, 491-525.
Mysen, B. O. (1979) Trace element partitioning between garnet peridotite minerals and water-rich vapour: experimental data from 5 to 30 kbar. Amer. Mineral., 64, 274-287.
Nakamura, N. (1974) Determination of REE, Ba, Fe, Mg, Na and K in carbonaceous and ordinary chondrites. Geochim. Cosmochim. Acta, 38, 757-775.
Nance, W. B. and Taylor, S. R. (1976) Rare earth element patterns and crustal evolution. I Australian post - Archaean sedimentary rocks. Geochim. Cosmochim. Acta, 40, 1539-1555.
Newton, R. C., Smith, J. V. and Windley, B. F. (1980) Carbonic metamorphism, granulites and crustal growth. Nature, 288, 45-50.
Peterman, Z. E. (1979) Strontium isotope geochemistry of late Archaean to late Cretaceous tonalites and trondhjemites. In: Barker, F.

(ed.), Trondhjemites, dacites and related rocks. Elsevier, Amsterdam, 122-147.

Pichamuthu, C. S. (1960) Charnockite in the making. Nature, 188, 135-136.

Pichamuthu, C. S. (1965) Regional metamorphism and charnockisation in Mysore State, India. Indian Mineral., 6, 119-226.

Piepgras, D. J. and Wasserburg, G. J. (1980) Neodymium isotopic variations in seawater. Earth Planet. Sci. Lett., 50, 128-138.

Pride, C. and Muecke, G. K. (1980) Rare earth element geochemistry of the Scourian Complex, NW Scotland - evidence for the granite granulite link. Contrib. Mineral. Petrol., 73, 403-412.

Ronov, A. B. and Yaroshevski, A. A. (1969) Chemical composition of the Earth's crust. In: Hart, P. J. (ed.), The Earth's crust and upper mantle. American Geophysical Union Monograph No. 13, American Geophysical Union, Washington, D.C., 37-57.

Rogers, N. W. and Hawkesworth, C. J. (1982) Proterozoic age and cumulate origin for granulite xenoliths, Lesotho. Nature, 299, 409-413.

Rogers, N. W. (1977) Granulite xenoliths from Lesotho kimberlites and the lower continental crust. Nature, 270, 681-684.

Saunders, A. D., Tarney, J. and Weaver, S. D. (1980) Transverse geochemical variations across the Antarctic Peninsula: implications for genesis of calc-alkaline magmas. Earth Planet. Sci. Lett., 46, 344-360.

Selverstone, J. (1982) Fluid inclusions as petrogenetic indicators in granulite xenoliths, Pali-Aike volcanic field, Chile. Contrib. Mineral. Petrol., 79, 28-36.

Shaw, D. M., Reilly, G. A., Muysson, J. R., Pattenden, G. E. and Campbell, F. E. (1967) An estimate of the chemical composition of the Canadian Precambrian Shield. Can. J. Earth Sci., 5, 829-853.

Sheraton, J. W., Skinner, A. C. and Tarney, J. (1973) The geochemistry of the Scourian gneisses of the Assynt district. In: Park, R. G. and Tarney, J. (eds.), The early Precambrian of Scotland and related rocks of Greenland. University of Keele, Keele, 13-30.

Sighinolfi, G. P., Figueredo, M. C. H., Fyfe, W. S., Kronberg, B. I. and Tanner Oliveira, M. A. F. (1981) Geochemistry and petrology of the Jequie granulitic complex (Brazil): an Archean basement complex. Contrib. Mineral. Petrol., 78, 263-271.

Smithson, S. B. (1978) Modelling continental crust: structural and chemical constraints. Geophys. Res. Lett., 5, 749-752.

Smithson, S. B., Brewer, J. A., Kaufman, S., Oliver, J. E. and Zawislak, R. L. (1980) Complex Archaean lower crustal structure revealed by COCORP crustal reflection profiling in the Wind River Range, Wyoming. Earth Planet. Sci. Lett., 46, 295-305.

Smithson, S. B. and Brown, S. K. (1977) A model for lower continental crust. Earth Planet. Sci. Lett., 35, 134-144.

Smithson, S. B. and Decker, E. R. (1974) A continental crustal model and its geothermal implications. Earth Planet. Sci. Lett., 22, 215-225.

Tarney, J. (1976) Geochemistry of Archaean high-grade gneisses, with implications as to the origin and evolution of the Precambrian crust. In: Windley, B. F. (ed.), The early history of the Earth. Wiley, London, 405-418.

Tarney, J., Skinner, A. C. and Sheraton, J. W. (1972) A geochemical comparison of major Archaean gneiss units from Northwest Scotland and East Greenland. Rep XXIV Int. Geol. Congr. Montreal, 1, 162-174.

Tarney, J. and Windley, B. F. (1977) Chemistry, thermal gradients and evolution of the lower continental crust. J. Geol. Soc. Lond., 134, 153-172.

Taylor, S. R. (1964) Abundance of elements in the continental crust: a new table. Geochim. Cosmochim. Acta, 28, 1273-1285.

Taylor, S. R. (1967) The origin and growth of continents. Tectonophysics, 4, 17-34.

Taylor, S. R. (1977) Island arc models and the composition of the continental crust. In: Talwani, M. and Pitman, W. C. (eds.), Island arcs, deep sea trenches and back-arc basins, Maurice Ewing series, No. 1. American Geophysical Union, Washington, D. C., 325-335.

Taylor, S. R. (1979) Chemical composition and evolution of the continental crust: the rare-earth element evidence. In: McEllinny, M. W. (ed.), The Earth: its origin, structure and evolution. Academic Press, London, 353-376.

Taylor, S. R. and McLennan, S. M. (1981) The composition and evolution of the continental crust: rare earth element evidence from sedimentary rocks. Phil. Trans. Roy. Soc., A301, 381-399.

Touret, J. (1971) Le facies granulite en Norwege Meridionale: II Les inclusions fluides. Lithos, 4, 423-436.

Touret, J (1974) Facies granulite et fluides carboniques. Bull. Ann. Soc. Geol. Belg., P. Michot vol., 267-287.

Veizer, J. and Compston, W. (1976) $^{87}Sr/^{86}Sr$ in Precambrian carbonates as an index of crustal evolution. Geochim. Cosmochim. Acta, 40, 905-914.

Vinogradov, A. P. (1962) Average contents of chemical elements in the principle types of igneous rocks of the Earth's crust. Geochemistry, 7, 641-664.

Weaver, B. L. (1980) Rare-earth element geochemistry of Madras granulites. Contrib. Mineral. Petrol., 72, 271-279.

Weaver, B. L. and Tarney, J. (1980a) Rare-Earth geochemistry of Lewisian granulite-facies gneisses, N.W. Scotland: implications for the petrogenesis of the Archaean lower continental crust. Earth Planet. Sci. Lett., 51, 279-296.

Weaver, B. L. and Tarney, J. (1980b) Continental crust composition and nature of the lower crust: constraints from the mantle Nd-Sr isotope correlation. Nature, 286, 342-346.

Weaver, B. L. and Tarney, J. (1981) Lewisian gneiss geochemistry and Archaean crustal development models. Earth Planet. Sci. Lett., 55, 171-180.

Weaver, B. L. and Tarney, J. (1982) Andesitic magmatism and continental growth. In: Thorpe, R. S. (ed.), Andesites. Wiley, London, 639-661.

Weaver, B. L. and Tarney, J. (1983) Elemental depletion in Archaean granulite-facies rocks. In: Atherton, M. P. and Gribble, C. D. (eds.), Migmatites, melting and metamorphism. Shiva Publ., Nantwich, 250-263.

Wildeman, T. R. and Haskin, L. A. (1973) Rare earths in Precambrian sediments. Geochim. Cosmochim Acta, 37, 419-438.

Wise, D. U. (1974) Continental margins, freeboard and volumes of continents and oceans through time. In: Burke, C. A. and Drake, C. L. (eds.), The geology of continental margins. Springer-Verlag, Berlin, 45-58.

Wood, D. A., Joron, J-L., Treuil, M., Norry, M. J. and Tarney, J. (1979) Elemental and Sr isotope variations in basic lavas from Iceland and the surrounding ocean floor. Contrib. Mineral. Petrol, 70, 319-339.

Wood, D. A. Tarney, J. and Weaver, B. L. (1981) Trace element variations in Atlantic Ocean basalts and Proterozoic dykes from northwest Scotland: their bearing upon the nature and geochemical evolution of the upper mantle. Tectonophysics, 75, 91-112.
Wyllie, P. J. (1977) Crustal anatexis: an experimental review. Tectonophysics, 43, 41-71.
Wyllie, P. J. (1982) Subduction products according to experimental prediction. Bull. Geol. Soc. Am., 93, 468-476.
Wyllie, P. J. and Sekine, T. (1982) The formation of mantle phlogopite in subduction zone hybridization. Contrib. Mineral. Petrol., 79, 375-380.
Zartman, R. E. and Doe, B. R. (1981) Plumbotectonics - the model. Tectonophysics, 75, 135-162.

EVOLUTION, GROWTH AND STABILIZATION OF THE PRECAMBRIAN LITHOSPHERE

Alfred Kröner

Institut für Geowissenschaften, Universität Mainz, Postfach 3980, 6500 Mainz, West Germany

ABSTRACT

Modern plate tectonic processes do not appear to be very efficient in continent-building since they have contributed little to the total volume of the continental lithosphere over the last 900 Ma. The much higher crust-formation rate in the Archaean cannot be ascribed to faster plate motion or frequent collision of many small plates during that time but appears to be related to extensive melting in the upper mantle and underplating of dense magmas that could not penetrate the overlying continental crust. These magmas provided a reservoir for most of the typical Archaean magmatic rock associations found in greenstone-grainite-gneiss terrains.

It is argued that the predominantly biomodal volcanic suites of Archaean greenstone belts and the genetically related tonalite-trondhjemite-granite plutons did not originate through growth on intra-oceanic island arcs or in Andean-type arcs and marginal basins but in intracontinental rift settings.

Magmatic underplating led to non-linear growth and differentiation of Archaean continental crust and is also seen as the prime cause for early lithospheric thickening and stabilization, allowing large intracontinental basins to form since 3.5 Ga ago. The end-Archaean global crust-forming event generated the first large cratons whose subsequent stability may be the result of subcrustal lithospheric thickening through melt extraction and depletion, underplating and carbonic granulite metamorphism in the lower crust.

Magmatic underplating continued to remain an important process in the generation of volcanic sequences in the Lower Proterozoic intracontinental basins although there appears to be first evidence for local horizontal crustal accretion in magmatic arcs at about 2 Ga ago. Most of the widespread Proterozoic mobile belts lack typical Wilson-cycle signatures and are interpreted by a model involving extensive crustal thinning, subcrustal lithosphere delamination and crust restacking during basin closure and orogeny. New and fertile, but initially mechanically weak lithosphere is accreted to the crust from below during this process. By the end of the Precambrian horizontal accretion of juvenile continental crust became widespread, but vertical accretion may still be important in Phanerozoic crust-forming events.

It is concluded that vertical rather than horizontal growth of lithosphere has dominated global evolution during the first 3.5 Ga of earth's history and that this mechanism also accounts for the observed magmatic rock assemblages as well as for lithospheric stabilization and cratonization of continental crust.

1. INTRODUCTION

Modern plate tectonic processes contribute to crustal growth (i.e. irreversible addition of juvenile, mantle-derived material) primarily by lateral accretion in

Andean- or Cordilleran-type continental margins and in island arcs. A lesser amount of juvenile crust is added by volcanic overplating accompanying intracontinental rifting and/or continental breakup (e.g. continental flood basalts in Southern Africa, Antarctica, India, Siberia) and through intraplate hot-spot volcanism. In all these settings there must also be some crustal addition through magmatic underplating, but the proportion of this contribution is still unknown.

Contemporary plate tectonics can now be traced back to at least the late Precambrian (Kröner, 1981a) and there is convincing evidence, for instance, that lateral crustal accretion of juvenile crust was the dominant process in the evolution of the Arabian-Nubian shiled between c. 900 Ma and 500 Ma ago (Greenwood et al., 1982). Further back in time the geological record that permits reconstruction of tectonic processes is much more fragmentary. This situation, in combination with the remarkable lack of some typical features characterizing modern accretionary terrains (e.g. ophiolites, blue-schists, melanges) in more ancient regions, has led to the present controversy that centers on the general question: "What was the dominant process of crust-formation in the Precambrian?"

There is abundant evidence for juvenile crustal addition throughout the Precambrian, and many rock types previously regarded to represent products of crustal reworking have since been shown by isotopic work to originate from mantle-derived magmas (e.g. Moorbath, 1978). However, following uniformitarian principles, these isotopic and geochemical arguments were used to infer contemporary plate tectonic settings for these suites (Windley, 1981) while geological field relationships often did not support such models (Kröner, 1981a).

Rock associations and deformational patterns surprisingly reminiscent of modern active continental margins have been recognized as far back in time as the middle Proterozoic (e.g. Wopmay Orogen, Canada, Hoffman, 1980; Svecofennian assemblage of Sweden and southwest Finland, Hietanen, 1975; Wilson, 1982). Nevertheless, the lack of ophiolites, accretionary melanges and blueschists in these terrains may indicate that the mechanism of crustal interaction differed from that presently observed (Kröner, 1981a) although some form of plate tectonics must have operated.

The above controversy also characterizes the discussion on Archaean crustal evolution where models favouring a dominance of vertical tectonics still abound, particularly in greenstone belt genesis (e.g. Anhaeusser, 1981). However, the importance of horizontal shortening is now increasingly recognized in both greenstone and gneiss terrains (e.g. Glover and Groves, 1981; Tankard et al., 1982), and new paleomagnetic evidence supports the independent motion of several continental plates since the early Archaean (Kröner and McWilliams, 1982). These results lead to the inescapable conclusion that plate motion has probably occurred since the first generation of rigid lithosphere. Therefore, the concept of lithospheric plates moving over an asthenosphere should be accepted by any model discussing continental growth and global tectonics through time.

Uniformitarian plate tectonics implies a more or less uniform growth of continental crust through time, and this concept was in apparent agreement with the model proposed by Hurley and Rand (1969). However, heat flow arguments contradict this evolution and demand high crust-generation rates in the early Precambrian (Bickle, 1978). This is now substantiated by geochemical considerations and isotopic constraints (e.g. McCulloch and Wasserburg, 1978; Jacobsen and Wasserbrug, 1979; McLennan and Taylor, 1982) suggesting that 70-85% of the present continental crust apparently formed in Archaean times between c. 3.7 and 2.5 Ga ago (Fig. 1). We have to examine whether this is the result of faster plate motion and, therefore, much shorter ocean crust recycling times, or whether we had many small plates, resulting in a much greater length in total plate margins, more frequent collisions and allowing higher crust-production rates, or whether we may invoke a different mechanism not resulting in lateral accretion of juvenile continental crust.

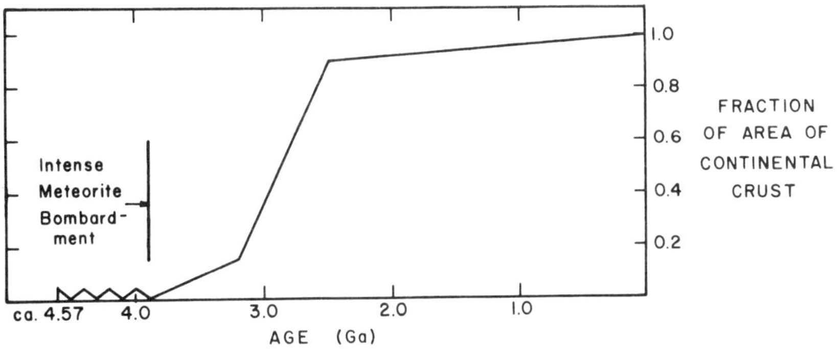

Fig. 1: Fraction of area of exposed continental crust as a function of age, suggested by trace element data and isotopic modelling. Note masive crustal growth between c. 3.2 Ga and 2.5 Ga (from McLennan and Taylor, 1982).

Whatever the mechanism in the Archaean, the above crust-formation model implies that little crust was added to the continents during the past 2.5 Ga. For the Proterozoic I had reached a similar conclusion from an examination of African mobile belts (Kröner, 1977) that indicated a predominance of crustal

reworking and intracontinental tectonics and a lack of plate margin activity. The growth model also implies that the present Wilson-cycle and associated subduction-related continental margin and island-arc magmatism do not appear to be very efficient continent-building mechanisms since they have contributed little to the total volume of the continental crust over the last 900 Ma.
Furthermore, it is of interest to examine the question why crustal growth took place episodically rather than continuously (Moorbath, 1978; De Paolo, 1981), and whether crustal accretion events led to stabilization and, therefore, cratonization of crust, or whether they generated zones of lithospheric weakness that eventually permitted continental breakup and dispersion.
Lastly, some speculation is warranted concerning the effect of lithospheric thickening through time on crustal accretion and stabilization.

2. PRECAMBRIAN PLATE MOTION

The widely held opinion that higher heat flow in the Precambrian resulted in faster plate motion (e.g. Bickle, 1978; Dewey and Windley, 1981) is based on the argument that the greater heat flux during those times must have been removed via the oceanic crust, either by an increased rate of plate production or by an increased total length of oceanic ridges (Bickle, 1978). In addition, it has been argued that plate motion would have been faster due to a lower viscosity in the mantle. For example, if the Archaean terrestrial heat flow was 2-3 times present rates (Lambert, 1981), upper mantle temperatures were likely to have been up to 150°C higher than now (Davies, 19790; Cook and Turcotte, 1981; Smith, 1981), as shown by the widespread occurrence of MgO-rich komatiitic volcanic rocks in Archaean terrains, that are commonly ascribed to high degrees of partial melting in the mantle (Green, 1975, 1981). Hager and O'Connell (1980) calculated that a temperature increase of 100°C in the mantle results in a viscosity decrease by a factor of 10, leading to a three-fold increase in plate velocities. This conclusion would allow for about 6 times the current oceanic crust formation rate of 3 km^2/a by 2.8 Ga ago or an average age of ocean crust at the time of subduction of only c. 20 Ma as demanded by the calculations of Sleep and Windley (1982).

However, presently available paleomagnetic data for Archaean rocks do not support this concept in that they fail to reveal significantly higher minimum continental velocities with respect to the poles in the early Precambrian. These data come from the Superior Province in Canada (Dunlop, 1981), the Pilbara Block of Western Australia (McElhinny and Senanayake, 1980) and from preliminary Archaean pole positions from the Kaapvaal craton of Southern Africa (Kröner and McWilliams, 1982).

The Canadian results imply an average velocity of at least 2 cm/a for the period 2.8 Ga to 2.5 Ga ago (Ullrich and Van der Voo, 1981), comparable to the Western Australian data that suggest a minimum drift rate of c. 1.5 cm/a for the period 3.5 to 2.4 Ga ago. The Kaapvaal craton appears to have experienced an appreciably higher minimum velocity of about 4 cm/a during the same period, but still comparable to present rates of lithospheric motion. The limited data do not yet permit recognition of strongly fluctuating drift rates in the Archaean as has been postulated for later times (Ullrich and Van der Voo, 1981). But an important conclusion to emerge is that although each Archaean craton has its own velocity signature, the overall mean minimum velocity of 2.8 cm/a for the continents in post-Archaean times (Ullrich and Van der Voo, 1981) does not appear to have been significantly higher in earlier periods as is suggested by the above results.

This is also borne out by the geologcial data. Dewey (1982) has argued that increased rates of plate motion are signified by marine transgressions, a higher incidence of ophiolite obduction and the generation of blueschists and granitoid rocks. Except for the granitoids, whose origin and role in Archaean crustal evolution will be discussed below, none of these processes are characteristic of the early earth. In fact, no ancient ophiolites or bluischists have been reported despite excellent preservation of Archaean upper crustal segments, and it remains one of the enigmatic problems for most uniformitarian models to explain this phenomenon if early Precambrian plates moved indeed much faster and subduction-complexes would thus have risen correspondingly fast to preserve primary high-pressure metamorphic assemblages. Also, if Archaean oceanic crust was non-subductible as is sometimes claimed (Green, 1975; Baer, 1977, 1981), because its turnover was fast and thus did not allow for substantial cooling (Bickle, 1978), why did this crust not preferentially obduct, particularly if there was less elevation contrast between surfaces of continents and the mean depth of the ocean floor (Dickinson, 1981; Dewey and Windley, 1981)?

In the present plate tectonic regime there is a close relationship between plate velocity (and, by implication, crust-generation rate) and the total length of active plate margins (Forsyth and Uyeda, 1975). If, in the Archaean, plates did not move significantly faster than those of today, it is also unlikely that the total length of Archaean active plate margins was appreciably greater than at present. It remains unexplained, therefore, how early Precambrian plate interaction, if it operated in the same way as today, could account for crustal growth rates greater than 6 km^3/a, or approximately 10 times present rates, as required by geochemical modelling (McLennan and Taylor, 1982) and the isotopic constraints mentioned earlier.

The conclusion seems inescapable, therefore, that most of the continental growth in early Precambrian time was not by lateral addition but through processes of vertical accretion that resulted in massive transfer of juvenile material from the mantle to the base of the crust and, through further magmatic differentiation, into higher levels of the crust. The mechanism and consequence of such "crustal underplating" are further explored below, and it is suggested that this process has played an important role in crustal stabilization and lithospheric thickening as has already been suggested by Jordan (1979, 1981).

3. EVOLUTIONOF ARCHAEAN GRANITE-GREENSTONE TERRAINS

Currently popular views on the evolution of granite-greenstone terrains are all compatible with the concept of moving plates in the Archaean, but differ substantially in tectonic development and in the generation and emplacement of magmatic rocks. These models have been described in detail elsewhere (e.g. Windley, 1976, 1977; Glover and Groves, 1981; Kröner, 1981b) and can be grouped into two categories: 1. Subduction-related magmatic arc or marginal basin settings; 2. Intracontinental rift settings.

The first category models consider processes in modern intra-oceanic or Andean-type magmatic arcs and genetically related marginal basins as the closest analogue to Archaean crust-forming mechanisms and are based largely on geochemical arguments (e.g. Hart et al., 1970; Jahn et al., 1974).

Condie (1980) suggested that the first segments of surviving continental crust evolved entirely from oceanic andesite arc systems at convergent plate boundaries between 4.0 and 3.5 Ga ago, while Andean-type and/or maginal basin evolution should have followed between 3.5 Ga and 2.7 Ga ago.

The assumption of an early intra-oceanic development is based on the belief that komatiitic to tholeiitic volcanic sequences in the lower parts of the oldest known greenstone belts that have no recognized basement represent primeval oceanic crust (e.g. Anhaeusser, 1981; Condie, 1981a). For the younger greenstone successions of the late Archaean that show abundant evidence for the presence of sialic material or that were found resting on continental crust, the continental margin or marginal basin model appeared more appropriate (e.g. Tarney et al., 1976; Windley, 1977; Condie, 1981a).

There are several serious objections to these views that require short discussion. First, the oldest known greenstone suites (e.g. Isua, Greenland; Dwalile, Swaziland; Barberton, South Africa; Selukwe, Zimbabwe; Pilbara, Western Australia) are all dominated by bimodal volcanic associations and contain no or only a small proportion of true andesites. This bimodal distribution pattern is also reflected in the composition of the grey gneiss complexes with which these greenstone belts are associated (Barker and Arth, 1976).

Second, there is growing evidence that virtually all the above sequences were deposited in shallow water as shown by the presence of silicified evaporites, stromatolite-bearing carbonates, BIF and subaerial volcanics (for review see Bickle and Briksson, 1982; Kroner, 1983b), and at least for Barberton and Pilbara these basins are interpreted to have evolved on older sialic crust and were marginal to major oceans (Eriksson, 1981). J. F. Wilson (pers. comm., 1981) has now documented the first direct evidence for sialic basement under the 3.4 Ga old Selukwe greenstones in Zimbabwe.

Third, one of the most surprising aspects of early greenstone belts is the excellent state of preservation, generally low degree of regional metamorphism and, locally, lack of significant penetrative deformation in the komatiite-tholeiite piles that have reported apparent thicknesses of 5-15 km (e.g. Viljoen and Viljoen, 1969). If these sequences indeed represent ancient oceanic crust, it is incomprehensible why such crust -- considering appreciably higher Archaean mantle and ocean temperatures than those of today (Fyfe, 1978) -- should be so well preserved, while superheated aqueous solutions have profoundly altered modern ocean crust to depth of more than 5 km (East Pacific Rise tudy Group, 1981), and the present oceanic temperature regime is consistent with steep geothermal gradients of 200° - 400°/km (Geary and Kay, 1982) and upper amphibolite facies metamorphism in layer 3 between 5 km and 12 km depth (Oskarsson et al., 1982).

It is feasible, therefore, to assume that the greenstone successions were laid down on cool crust that effectively isolated the depositional environment from the hotter mantle, and that near-surface magma chambers and associated vigorous aqueous convection were not responsible for the production of these volcanic rocks. The lack of layered crust with sheeted dykes in Archaean greenstone suites further supports this contention.

Lastly, the use of geochemical data and discrimination diagrams to infer ancient tectonic settings must be viewed with considerable caution. It has been demonstrated that low temperature alteration and crustal contamination can obliterate primary magmatic trends and may impose an apparent calc-alkaline element distribution pattern on originally tholeiitic rocks (for review see Kröner, 1983b). DePaolo (1982) has demonstrated that even the rare-earth elements, commonly regarded as immobile, can be strongly affected by these processes. Also, virtually all discrimination methods are based on modern and "normal" rock compositions for the respective tectonic settings and fail to identify the correct environment for volcanic rocks with unusual compositions such as, for example, ocean floor andesites (Hekinian, 1974; Thompson et al., 1978) and anomalous ridge basalts (Prestvik, 1982). The inherent weakness of most arguments that apply present-day geochemical-tectonic correlations to early Precambrian times results from an insufficient understanding of geochemical behaviours of many trace elements during partial melting and crystal fractionation under different P, T, X conditions (Jahn and Sun, 1979) and is also due to the fact that the geochemistry and evolution of several rock types such as, for example, komatiites (Arndt and Nisbet, 1982), tonalites (Barker et al., 1981) and siliceous volcanics (Condie, 1976) are distinctly different from those formed in post-Mesozoic times.

Although present dogma holds that calc-alkaline magmas are predominantly produced in the mantle wedge overlying subduction zones (Gill, 1981; Thorpe, 1982), there are examples of anorogenic basalt - tholeiitic andesite - dacite - rhyolite associations from the Tertiary (Wilkinson and Binns, 1977) to the Archaean (Hegner et al., 1981), and these assemblages apparently occur in extensional environments where crustal melting is required to account for the observed chemical variations. Also, Giles and Hallberg (1982) described an Archaean calc-alkaline suite from the Yilgarn Block of Western Australia, identified as the remnant of a stratovolcano and completely surrounded by a "sea" of genetically unrelated basaltic rocks. They state that this relationship is difficult to reconcile with subduction, even on a small scale.

Several recent investigations have shown that andesitic to rhyolitic rock types in extensional settings may be produced by varying degrees of mixing lf lower continental crust and mantle material (for summaries see EOS, 61, 67-68, 1980) and that underplating of dense picritic to gabbroic magmas at the base of the crust played an important role in this process (Cox, 1980; Ewart et al., 1980; see Fig. 2).

Recent geochemical data also contradict the earlier view of ensimatic greenstone belt evolution since neither the tholeiitic nor the calk-alkaline suites are strictly identical to modern ocean floor and island arc volcanics (e.g. Hawkesworth and O'Nions, 1977; Glikson, 1979; Condie, 1981a; Grachev and Fedorowsky, 1981). The preferred occurrence of MgO-rich volcanics in the lower parts of greenstone assemblages, as also found in Phanerozoic rift environments (Hawkesworth and O'Nions, 1977), the pronounced biomodality of many greenstone successions and the close spatial relationship of greenstones with granitoid crust all favour a rift-related setting for the initial stage of greenstone belt formation.

Many of the objections enumerated above can also be raised against the marginal basin model as proposed by Tarney et al. (1976). This setting fails to account for many of the observed greenstone belt relationships (e.g. Groves et al., 1978; Platt, 1980; Hickman, 1981; Gee et al., 1981; Archibald et al., 1981), in particular the apparently large size of greenstone basins and the remarkably continuous stratigraphy in some regions. Furthermore, available age data suggest that most greenstone belts and associated granitoids had evolutionary histories extending over c. 200 Ma or more while, on the average, recent marginal basins have relatively short lifetimes of 40 Ma or less (Talwani and Langseth, 1981).

To summarize, the intraoceanic and island-arc-related greenstone models are in conflict with greenstone belt stratigraphies, geological field relationships and with geochemical criteria (Kröner, 1983b), and the data appear more compatible with rifting environments.

Continental rift models have been discussed at length in recent papers (for reviews see Glover and Groves, 1981; Kröner, 1981a, 1983b; Bickle and Eriksson, 1982), and the suggested scenarios range from intracontinental grabens where thickened and stabilized older lithosphere inhibited proper greenstone basin evolution ("failed greenstone belt", Hegner et al., 1981) through pull-apart basins, deep rifts and small ocean basins of Red-Sea type (Kröner, 1981a, 1983b).

Bickle and Eriksson (1982) have related the evolution of several such Archaean basins to the two-stage subsidence model of McKenzie (1978), that infers lithospheric stretching and thinning during the rifting phase. They show that subsidence in greenstone basins cannot be due to loading by the dense volcano-sedimentary pile since virtually all rocks were deposited in shallow-water. Instead, it is likely that crustal thinning produced listric fault patterns in the upper crust, similar to those observed in the Basin-and-Range Province of the western USA (Fig. 3), while the lower crust yielded by laminar flow. Such basin widening mechanism could explain the apparent great thickness in greenstone successions.

Much of the early deformation seen in greenstone belts is probably not due to diapiric rise of tonalitic plutons but relates to considerable horizontal translation, including nappe formation, during basin closure (e.g. Stowe, 1974; Platt, 1980; De Wit, 1982; Jackson and Robertson, 1983). Many of these originally flat structures were later modified and frequently steepened by listric faulting or during the emplacement of the granitoid batholiths.

Detailed models for greenstone belt evolution based on the continental rift setting have recently been described by Condie (1981a), Goodwin (1981), Kröner

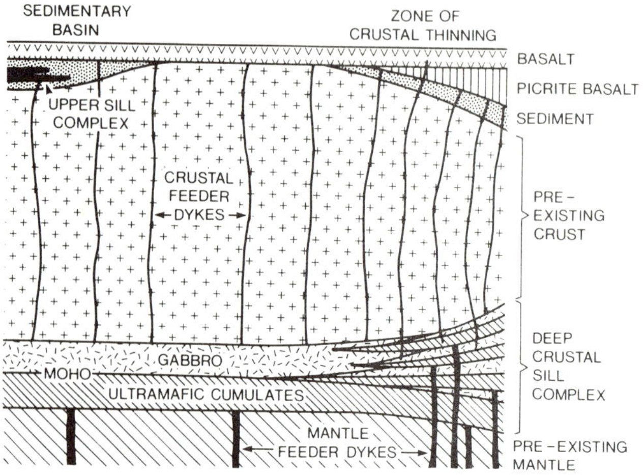

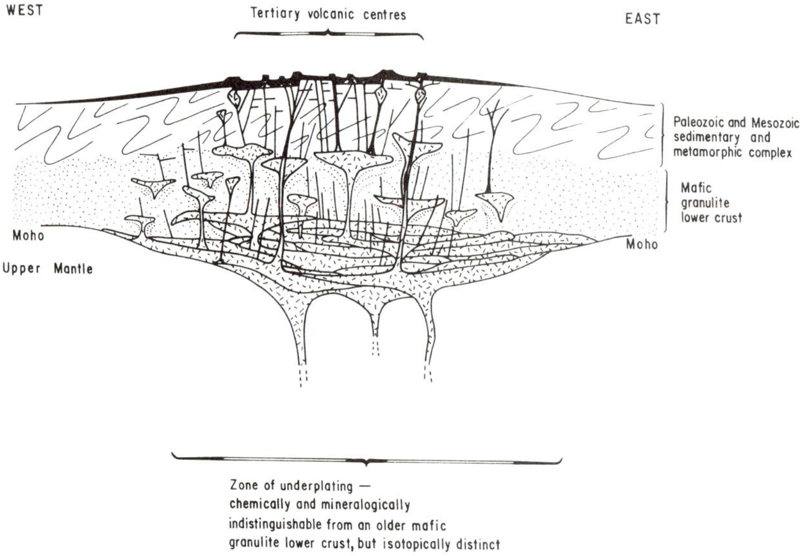

Fig. 2: Two models suggesting evolution of bimodal volcanic associations in continental rift environments through magmatic underplating.
top: Origin of Karroo volcanics in southern Africa (from Cox, 1980)
bottom: Origin of Tertiary lavas in Queensland, Australia (from Ewart et al. (1980).

(1981a, 1983b) and Lambert (1981), and Fig. 4 depicts several hypothetical stages from greenstone basin to deformed belt.

The fundamental difference between the rift and magmatic arc models from the point of view of crustal growth is that the former requires massive vertical accretion while the latter implies horizontal accretion through arc magmatism and subsequent tectonism. While the mechanism of crustal growth along modern active plate margins is reasonably well understood, there is much less agreement on possible processes that may generate the observed magmatic rock

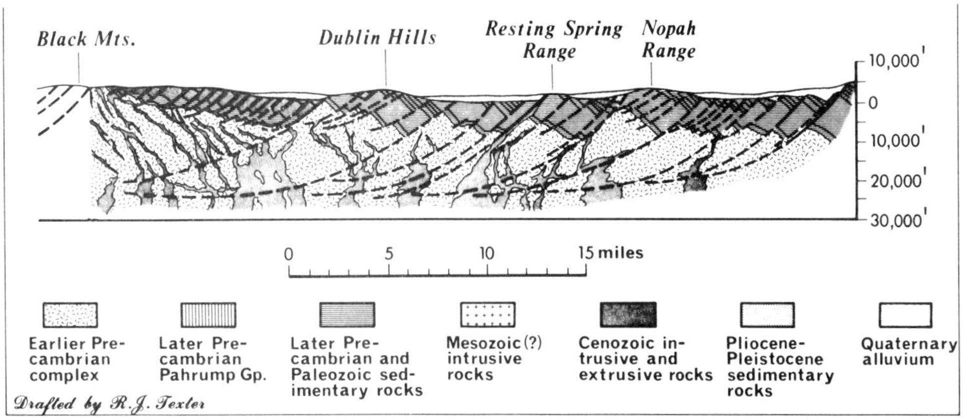

Figure 3: Geologic cross section through central and southern Death Valley, Southwestern Great Basin, Basin and Range Province, USA, showing suggested basin widening through upper crustal listric faulting and lower crustal flow (after Wright and Troxel, 1973).

associations in Archaean granite-greenstone terrains, and these are briefly discussed below.

4. SOURCE OF ARCHAEAN GREENSTONE BELT VOLCANIC ROCKS

As mentioned earlier, mafic underplating has recently been proposed as a likely mechanism to explain the source and evolution of Phanerozoic bimodal volcanic rocks in extensional environments. In the Archaean such underplating must have occurred on a much more extensive scale, since Hanson (1981) and Phinney et al. (1981) pointed out that mantle-derived mafic melts high in MgO and FeO such as komatiites have densities higher than most crustal rocks, and many such melts would therefore not reach upper crustal levels. Given the generally accepted high degrees of partial melting in komatiite source regions and melt densities in excess of 3 g/cm^3 (Nisbet, 1982), much of this material would be trapped at or near the base of the crust.

Hanson (1981) further suggested that such trapped melts would tend to form immiscible liquids, separate gravitationally, and then ascend into the crust to form either basalt and/or mafic tonalite or to solidify as dacite/rhyolite and/or granite. Phinney et al. (1981) and Ashwal et al. (1982) have proposed that, because of the long cooling times for underplated melts, substantial volumes of relatively light plagioclase would have formed that, at various stages of fractionation, could have produced anorthosite complexes in the lower crust.

The models of Cox (1980) and Ewart et al. (1980) likewise see the primary source for bimodal rift-generated volcanics in underplated mafic magmas, and there is also a strong likelihood of melting and mixing with components of the lower continental crust during melt ascent. In accounting for the origin of the

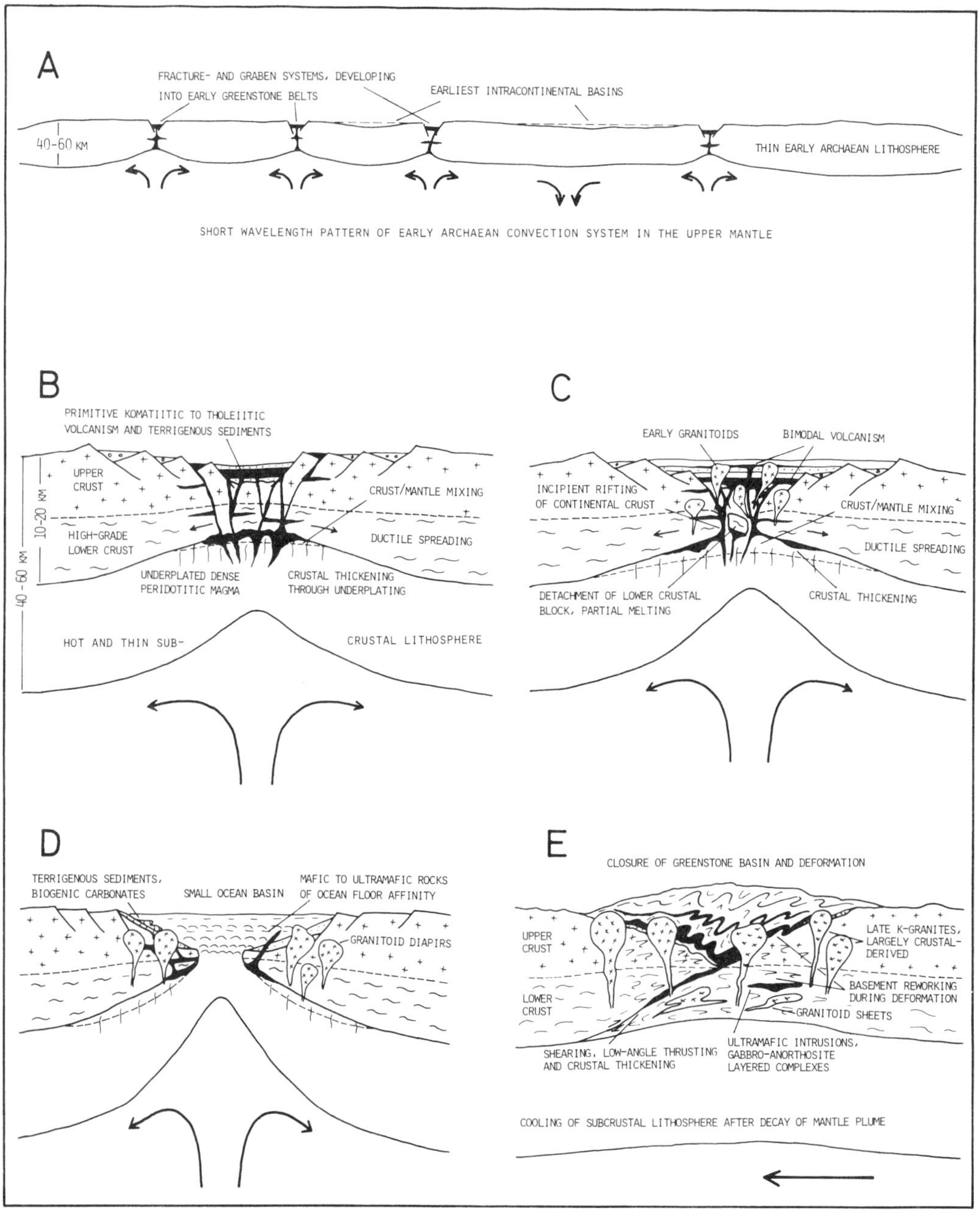

Fig. 4: Schematic cross sections showing suggested evolution of early Archaean lithosphere in response to small-scale convection pattern in the upper mantle (A) and various stages in the development of a greenstone belt (B-D). From Kröner (1983b).

clearly ensialic metavolcanics of the Archaean Yellowknife Supergroup in the Canadian Slave Province, Baragar (1966), for example, concluded that mafic magmas generated at the base of the crust were contaminated by crustal melts before they could be extruded, resulting in strong calc-alkaline trends superimposed on originally basaltic magmas. Significantly, the setting, as revealed by excellent exposures and detailed fieldwork, was one of stretching and rifting in the 3 Ga old pre-Yellowknife granitic crust (Padgham, 1981).

Early models for Archaean crustal evolution assumed that greenstone belts follow a pattern from ultramafic to mafic to siliceous volcanics, succeeded by a suite of granitoids evolving from Na-rich to K-rich types, and involving large-scale anatexis of older greenstone material. This process should reflect "maturing" of intraoceanic arcs to microcontinents (e.g. Glikson, 1976). However, the discoveries of felsic volcanic associations that occur in the lower parts of greenstone stratigraphies and that may be overlain by, or intercalated with, komatiitic lavas as well as the close spatial relationship between tholeiitic and calc-alkaline volcanics in individual belts (for details and references see Kröner, 1983b) cast strong doubt on the "growing arc" models. Furthermore, careful modern dating has demonstrated that both extrusive and granitoid intrusive rocks are broadly contemporaneous (e.g. Barton, 1981; DeLaeter et al., 1981; Scarer and Allegre, 1982), thereby implying that the latter are unlikely to represent melting products of the former. As shown below, large-scale melting of lower greenstone units to form granitoids is an implausible mechanism, and Bettenay et al. (1981) succinctly demonstrate that greenstone belts are probably shallow structures restricted to upper crustal layers.

Accepting the scenario of fracturing of continental crust during continental breakup or for the formation of pull-apart basins, and assuming that mafic underplating is an important process in this setting, the origin of Archaean bimodal volcanic suites may be broadly compared with the evolution of magmas in modern rift environments. Here both juvenile, mantle-derived melts and anatectic melts derived from partial melting of previously solidified mafic crust play a crucial role in the petrogenesis of observed bimodal associations (e.g. Oskarsson et al., 1982).

5. SOURCE OF ARCHAEAN BIMODAL GNEISS AND GRANITOID SUITES

All presently known Archaean terrain consist of varying proportions of greenstone belts, upper crustal granitoid intrusives and a variety of banded medium-to high-grade "grey gneisses" that frequently contain metasedimentary remnants and that are either older or formed broadly contemporaneously with the greenstone belt rocks (for summaries see Windley, 1977 and Glover and Groves, 1981). Taken together, an Archaean crustal column composed of these units would have broadly andesitic composition (Taylor, 1967) while the individual rock units themselves are remarkably andesite-free and almost universally reveal a pronounced "Daly gap" (Barker and Arth, 1976). The volumetrically most important granitoid rocks of most Archaean terrains are tonalites and cogenetic trondhjemites as well as their tectonized gneissic equivalents. Much controversy exists as to the origin of these rocks that typically form large composite batholiths and that often contain xenoliths of greenstone belt material as well as banded gneisses and/or migmatites that may represent either older sialic basement or products of interaction between greenstones and granitoid melt.

Several authors (for references see Kröner 1981a and 1983b) have suggested that the tonalite-trondhjemite complexes are derived from anatexis of a primitive mafic-ultramafic crust as found in lower greenstone successions and that the evolving magmas ascend to higher crustal levels where they intrude the higher greenstone units and form multilobate diapirs. This results in extensive assimilation of greenstone material and in the formation of marginal migmatites that develop into banded gneisses in high strain zones.

Petrogenetic considerations do not favour this model (Condie and Hunter, 1976; Hanson, 1981), and it is particularly difficult to explain the derivation of the widespread interbanded bimodal basalt-trondhjemite gneisses from such a process.

An alternative origin is proposed by most uniformitarian models that see similarities between Archaean tonalites and modern calc-alkaline batholiths and their gneissic roots that characterize Andean and Cordilleran belts (Windley, 1979; Hamilton, 1981; Weaver and Tarney, 1982). Tonalites and trondhjemites are generated from upper-mantle derived melts that originate above subduction zones and either intrude into the evolving arc or continental margin or accrete at or near the base of the crust (Weaver and Tarney, 1982) where they solidify and develop into granulite-gneiss belts (Windley, 1979). These processes require no or only short crustal residence times for the granitoid precursors - a feature supported by many isotope data (e.g. Moorbath, 1978).

The great volumes of tonalite-trondhjemite produced in the Archaean are conveniently ascribed to rapid ocean crust recycling and the frequent collision of numerous small, fast-moving plates (Dewey and Windley, 1981), but this scenario neither accounts for the spatial and temporal distribution of these rocks nor for their generation during worldwide well-defined crustal accretion events (Moorbath, 1978). The conflict of this hypothesis with the available paleomagnetic data has already been discussed.

The equation of ancient high-grade bimodal gneisses and their associated metasediments with modern accretionary thrust stacks that were transferred to deep crustal levels during the accretion event (Dewey and Windley, 1981) completely ignores the fact that the Archaean granulites frequently contain metasedimentary assemblages derived from shallow-water sandstone-shale-limestone sequences (Katz, 1977). The extensive granulite terrains of southern India, Enderby Land and southern Africa are also contrary to the narrow, short, linear belts that would be expected from subduction-accretion along small continental plates. An alternative origin for the Archaean high-grade gneisses through such processes as plane-strain thrust restacking of previously thinned or mechanically weakened continental crust (Myers, 1976; Kröner, 1983b) is more in line with field relationships and petrological data (e.g. Newton and Hanson, 1983).

A third origin for the tonalite-trondhjemite suite is proposed by Barker and Arth (1976), Barker et al. (1981), Condie and Hunter (1976), Condie (1981b), and Hanson (1981) on the basis of petrogenetic and geochemical modelling. These models show that K-poor granitoid magmas may be derived from mafic-ultramafic sources, either as products of partial melting of amphibolite or eclogite, or through fractional crystallizaton of basaltic liquid. The scarcity of intermediate magmas and melting experiments that yielded trondhjemite from an amphibolite source (Helz, 1976) favour the partial melting mechanism.

Partial melting of 10-20% amphibolite of basaltic to komatiitic composition at or near the amphibolite-granulite transition, i.e. near the base of the crust, produces tonalite-trondhjemite melts (Barker et al., 1981). In the lower crust these magmas intrude as sheets and, together with their host rocks, they are involved in plastic flow by lateral extension and form high-grade gneisses with generally flat foliations (Gastil, 1979). The 3.8 Ga old Sand river gneisses of the Limpopo belt (Fripp, 1981) may be an example of this model. At higher levels the melts may still intrude as sheets and form complex structural relationships with the tectonized roots of greenstone belts, including marginal zones of migmatites (author's observations in the Barberton region and in Swaziland, southern Africa; see also McGregor, 1979; Betenay et al., 1981). But most of the magma rises diapirically to form large composite batholiths that often engulf the previously deformed greenstones and obliterate their relationships with older basement.

The existence of significant amounts of trondhjemitic intrusives in modern extensional environments such as Iceland (Sigurdsson and Sparks, 1981), that were derived from basaltic sources, lends weight to the above model and suggests that Archaean hot spot activity may have been the primary cause for bimodal magmatic suites (e.g. Condie, 1975; Lambert, 1981).

The tonalite-trondhjemite suites with "primitive" isotope characteristics also satisfy the above model but do not preclude the possibility that their source terrains had very low Rb/Sr so that considerable crustal residence times up to several hundred million years are possible for the granitoid precursors (e.g. Barton, 1981; Hamilton et al., 1979). Jahn et al. (1982), for example, have demonstrated that tonalitic gneisses from the Archaean of northern Finland that yielded an emplacement age of 3.06-3.1 Ga have Sm-Nd isotopic systematics and REE patterns that suggest separation of the basaltic gneiss protolith from the upper mantle c. 3.35 to 3.53 Ga ago. This mafic material resided in the lower crust for some 300-600 Ma before partially melting to produce the tonalitic magma (Fig. 5).

In summary, Archaean tonalite-trondhjemite suites and compositionally similar grey gneiss terrains are unlikely to represent the products of anatexis of supracrustal greenstone belt lithologies. From the many dissimilarities of these rocks to Cordilleran batholiths and their root zones, their spatial distribution and their structural evolution, it is also considered unlikely that they formed along modern-type convergent zones. Models that consider underplating of dense mafic magmas in extensional environments, generation of tonalite-diorite-trondhjemite plutons from these sources and from remelting of older sialic gneissic crust are considered more realistic and satisfy constraints from field relationships, petrology and geochemistry.

6. ARCHAEAN LITHOSPHERIC THICKNESS AND CRATONIZATION

The preservation of Archaean lowe crustal assemblages with mineral parageneses suggesting normal averge geothermal gradients despite an appreciably higher global heat flow at that time has led to the suggestion that the ancient oceanic lithosphere was thin (20-30 km) whereas continental lithosphere under the cratons must have been fairly thick, thus insulating the continental crust against heat flux from the convecting mantle (Davies, 1979). By analogy, Wells (1981) has suggested that the survival of 30-40 km thick Archaean crust implies the presence of a stable subcrustal root zone to continents, not less than 200 km thick, that probably grew during crustal formation and thickening. These concepts are supported by heat flow data from Archaean cratons (Oxburgh, 1981), by seismic ScS wave travel-time variations (Jordan, 1979, 1981) and by geotherms deduced from kimberlite pyroxenes that require a plate thickness of at least 200 km (Davies and Strebeck, 1982). That such thick lithosphere was already present in late Archaean times is substantiated by the occurrence of kimberlite-derived diamonds in the c. 2.6 Ga old Witwatersrand conglomerates of South Africa (Hallbauer et al., 1980).

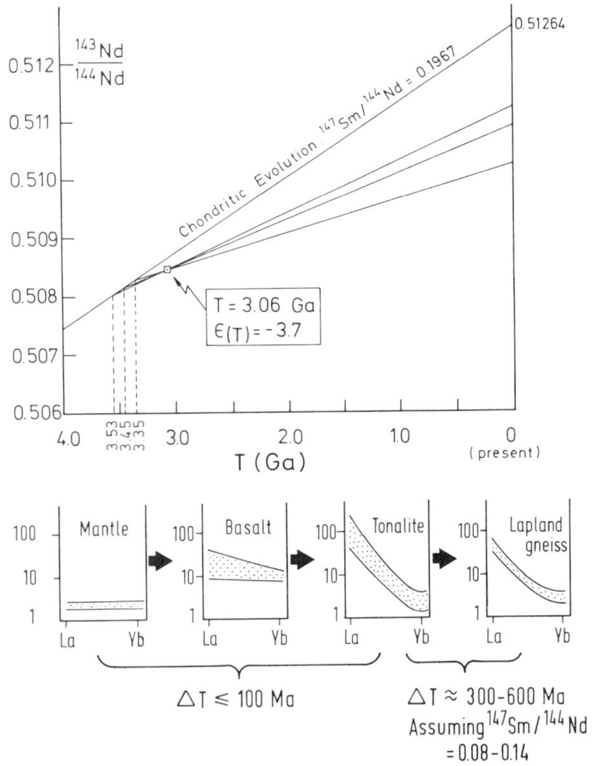

Fig. 5: Nd isotopic evolution diagram (top) and hypothetical 3-stage REE evolution model (bottom) for tonalitic gneisses of northern Finland (from Jahn et al., 1984).

Indirect evidence for a lithosphere of considerable thickness and stability and supporting continental crust of almost continental dimensions in the Archaean comes from greenstone belt lithologies and greenstone basin reconstructions in Western Australia (Fig. 6). The Pilbara greenstones were deposited on stable (presumably sialic) crust some 3.5 Ga ago in a basin almost 500 km wide (Hickman, 1981; Bickle and Eriksson, 1982), while the late Archaean greenstone basins in the Yilgarn Block were of even larger dimensions (Gee et al., 1981) and evolved on granitoid crust at leat 30 km thick (Archibald et al., 1981).

Jordan has argued repeatedly (1979, 1981) that subcrustal lithospheric growth results from basalt extraction in fertile, undepleted mantle, leaving a periodtitic residue that is less dense than the starting material. This depleted, light residue would therefore tend to float on undepleted mantle and would tend to be incorporated in the overlying lithospheric plate. Repeated continental basaltic volcanism would constitute primary episodes of basalt extraction from below, resulting in a relatively buoyant, thick subcontinental root that is not likely to be the site of further large-scale melting. An important implicaton of this persuasive model is that it explains cratonization of continental crust as a

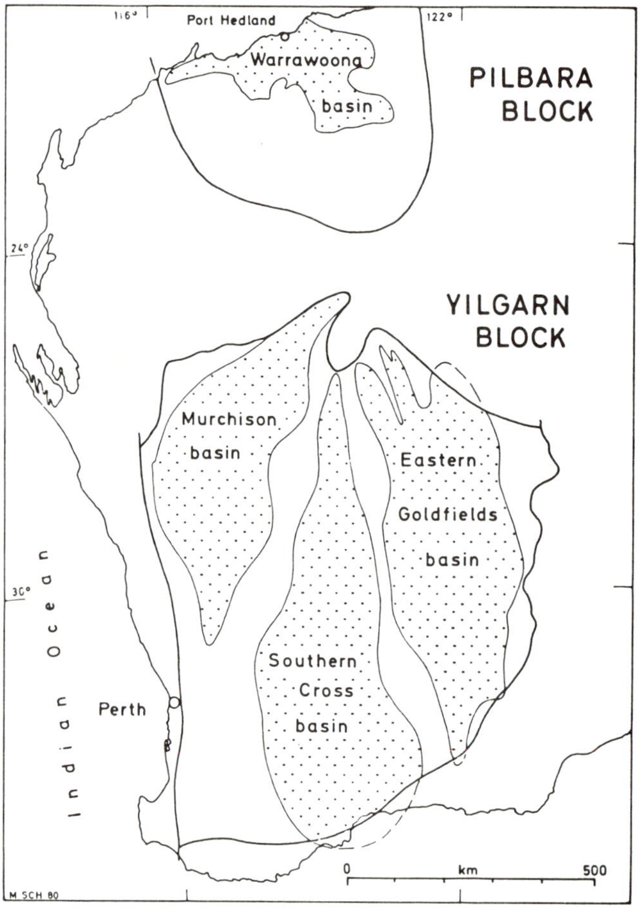

Fig. 6: Simplified map of Western Australia showing approximate size of Archaean greenstone basins in the Pilbara and Yilgarn blocks prior to their deformation and disruption through granitoid intrusions (from Kröner, 1981c).

direct consequence of early formation of a thick lithosphere (Jordan, 1981).

This vertical lithospheric growth model does not necessarily depend on the operation of continental margin subduction processes in the Archaean but accounts for crustal stabilization through widespread and extensive basalt extraction from the mantle. Such extraction may be seen in the form of massive crustal underplating as discussed in previous sections and the generation of greenstone belt volcanics as well as granitoid intrusives from this underplated reservoir and from the mantle below. The areally extensive worldwide crustal accretion event at the end of the Archaean has almost certainly resulted in substantial cratonization as revealed by thick, late Archaean to early Proterozoic intracontinental volcano-sedimentary successions some of which may constitute "failed" greenstone belts (Hegner et al., 1981).

The "subcontinental lithospheric root" concept may also be able to explain the observation that, in some cases, isotopic heterogeneities have persisted in local closed systems in the upper mantle under old cratons since Archaean times (Kramers, 1979) since this material has obviously not participated in later reequilibration with undepleted mantle through convection or significant melting.

There are several ways in which Archaean lithospheric growth by underplating can be envisaged. The Jordan (1979, 1981) model requires a heat source in the mantle to cause melting and basalt extraction. Under continental (and modern oceanic) areas such sources are generally seen in hot-spots or mantle plumes (e.g. Hofmann and White, 1982) and, in some cases, these thermal anomalies must have been of considerable dimensions as urged by the voluminous basalt suites such as the early Proterozoic Ventersdorp and Fortescue Supergroups of South Africa and Western Australia or by the Mesozoic Karroo volcanism of Gondwanaland.

Since mantle plumes are considered to be major agents for transferring heat from the deeper mantle (Morgan, 1971), they must have been more common in the Archaean than today (Condie, 1975), (Lambert, 1981), perhaps forming planar arrays (Hanson, 1981). They may ultimately have been responsible for greenstone belt formation and granitoid magmatism on and in the continental crust while the subcrustal lithosphere gradually (or episodically) thickened and thus contributed to continental stabilization.

An alternative mechanism to cause lithospheric underplating may be seen in the proposals of Hofmann and White (1982) and Ringwood (1982) whereby subducted oceanic lithosphere sinks into the mantle to depth of about 600 km where the basaltic component is denser than the surrounding mantle and thus keeps sinking while the former harzburgite part of the slab becomes buoyant and rises diapirically. This material, being less dense than undepleted mantle, just as in the Jordan model, underplates the lithosphere and causes thickening. In this model old cratons have thick lithosphere simply because they have been subjected to those type of underplating for long times.

Following a suggestion of Hofmann and White (1982), Ringwood (1982) further proposes that the process from subduction to thermal equilibration with the mantle and subsequent diapirism and underplating takes place on a time scale of 1-2 Ga. For the Archaean this may mean that the extensive global magmatism and continental stabilization at c. 2.7-2.5 Ga ago may be due to massive underplating as a result of substantial ocean crust recycling 1-2 Ga earlier, i.e. at a time when most models predict that continental crust formation began.

Early stabilization of continental crust, however, may not only depend on lithospheric thickness but also on the effective removal of heat-producing elements from the lower crust through regional metamorphism (Pollack, 1982). Many of the stabilized Archaean granite-gneiss-greenstone terrains are underlain by dense granulite material that shows the characteristic depletion of alkalies and certain other trace elements (e.g. Heier and Thorensen, 1971; Hart et al., 1981). It has been shown recently that this depletion is not due to magmatic differentiation during increasing P, T conditions, but that granulitization may occur without a significant change in temperature and pressure but merely as a result of dehydration effected through a CO_2-rich vapour phase that migrates upwards from below (carbonic metamorphism of Newton et al., 1980). Interestingly, the regional source of CO_2 is seen in massive basalt underplating (Newton et al., 1980).

Carbonic dehydration and depletion may have effectively contributed to stabilization of the crust. For example, devolatilized Archaean granulite terrains now exposed display little evidence of later deformation while those terrains where carbonic metamorphism did apparently not occur and where undepleted granulites are found as, for example, in the Jequie complex of Brazil (Sighinolfi et al., 1981), post-Archaean reworking has taken place (Cordani and Yier, 1978).

It is therefore possible that both subcrustal lithospheric thickening and lower crustal carbonic metamorphim are required to generate stable cratons, i.e. devolatilization and upward-motion of heat-producing elements appear to be decisive processes in lithospheric stabilization (Pollack, 1982) and are therefore likely to have operated extensively in the formation of large, stable continental plates that existed at the end of the Archaean (Kröner, 1983b).

In summary, thick continental roots under old shields may have formed early in the Earth's history and may have been responsible for the preservation of ancient lower continental crust. Together with devolatilization this may have been the prime reason for cratonization. Current models in support of this concept see the cause of thick subcontinental lithosphere in magmatic underplating and the operation of mantle-plumes.

7. EARLY PROTEROZIOC TECTONICS AND CRUSTAL ACCRETION

A period of comparative continental stability, following late Archaean cratonization through voluminous granitoid magmatism, crustal differentiation and lower crustal dehydration, characterizes most old shields in the early Proterozoic. In southern Africa considerable crustal stability was already achieved some 3 Ga ago as documented by still little-deformed volcanics and sediments that were deposited in large intracratonic basins (Kröner, 1981a, c; Bickle and Eriksson, 1982). Elsewhere this stability is documented at about 2.5-2.4 Ga ago by such sequences as the Fortescue in Western Australia and the Dharwar in Peninsular India. All these units developed in continental rift environments on thick, older granitoid crust and their subsequent deformation, if any, reflects the degree of stability of the underlying craton. In some cases the basin evolution closely follows the greenstone belt pattern of the Archaean but ultramafic volcanics and granitoids are less frequent and deformation is markedly less intense. The 3 Ga old Pongola sequence of southern Africa and the 2.4-2.5 Ga old Dharwar suites are examples of this kind and have been termed "failed greenstone belts" by Hegner et al. (1981) since they may reflect abortive attempts of the stabilized crust to break apart.

Pollack (1982) has suggested that lithospheric stability is an acquired property that requires an equilibration period of several hundred million years following a tectono-thermal event, during which the affected lithosphere is vulnerable to recurrent tectonism. The above examples demonstrate, however, that stabilization may occur over significantly shorter time scales and may be more dependent on effective lithospheric differentiation than on thermal equilibration. McLennan et al. (1983), for example, have shown that sedimentary REE patterns through time are fingerprints of crustal differentiation, and they demonstrated that the c. 3 Ga old cratonic Pongola Group sediments were already deposited on mature, stable K-feldspar-dominated crust of the Kaapvaal craton while, at the same time, sediments in other terrains reveal less advanced crustal differentiation and, correspondingly, do not reflect crustal stability in the depositional environment.

Condie (1982) has reviewed the composition of early and middle Proterozoic successions and concludes that the period before c. 1.7 Ga was dominated by stable continental margin or intracratonic basins. Many of the latter basins contain mafic volcanic rocks that reach several thousand meters in thickness as, for example, the c. 2.5 Ga old Ventersdorp Group in South Africa (Winter, 1976) and the equivalent Fortescue Group of Western Austrlia (Horwitz and Smith, 1978), and these rocks were undoubtedly produced in extensional intracratonic environments, presumably from underplated mantle sources.

It is important to note here that the virtually undeformed Ventersdorp Group contains basal komatiites, followed by tholeiitic basalt and basaltic andesite, all resting unconformably on continental crust and displaying a chemistry that is comparable to the volcanic sequences in Archaean greenstone belts (e.g. McIver et al., 1982). This situation again supports the rift setting for greenstone evolution and also indicates that vertical accretion through underplating extended well into the Proterozoic.

In regions where Archaean evolution did not generate thick subcrustal lithospheres and where early lower crustal dehydration did not occur, a variety of mixed volcano-sedimentary sequences were deposited on less stable crust and formed successions that are somewhat intermediate between Archaean greenstone suites and deposits in Phanerozoic "geosynclines" in that the proportion of more differentiated volcanics including andesites as well as immature sediments is appreciably higher than in most Archaean suites (e.g. Condie, 1982). The Birrimian of West Africa, the Jacobina and Minas Groups of Brazil and the Huronian of the Canadian shield may be examples of this type.

Although clearly associated with older continental crust the evolution of these sequences is still a matter of considerable debate since they were deposited in elongate, well-defined basins and experienced a tectonic history that reveals considerable horizontal shortening, involving both the basin fill and its basement. The Brazilian and the West African Birrimian sequences have been interpreted as greenstone belts, deformed between 2.4 and 2.0 Ga ago, and their evolution from rift, pull-apart or large-scale intracontinental transcurrent shear structures is compatible with the geological field relationships.

Other assemblages suggest still more mobile environments with frequent continental rifting and the local onset of buoyancy-driven subduction. Suggestions for middle Proterozoic convergent plate boundary tectonics come from the Wopmay orogen of the Canadian shield, the Svecofennides of the Baltic shield and the Mid-continent region of North America (for reviews see Medaris, 1983), while contrary, but less convincing data suggest the existence of a stable supercontinent throughout the Proterozoic (e.g. Piper, 1982).

An interesting setting has been described from the mid-Proterozoic Cape Smith foldbelt, New Quebec, Canada, which is part of the Circum-Superior belt system. Here, komatiitic basalt and low-Ti tholeiite have been interpreted as oceanic crust that should have developed in a small ocean following crustal thinning and subsequent continental rifting. The absence of subduction-related volcanics, however, and the geological similarities of the two rifted continental margins, suggest that the foldbelt does not constitute a major suture zone but reflects closure of a small basin of Red-Sea type (Hynes and Francis, 1982).

Whatever the precise mechanism in the above cases, it seems clear that the evolution of the early Proterozoic continental crust was dominated by intracrustal tectonic and magmatic activity (Rutland, 1982) that resulted in some addition to the continents by means of volcanic "overplating" while, at the same time, the lithosphere grew vertically as a result of cooling and magmatic differentiation. Lateral crustal accretion may have occurred in places but was probably less significant.

The similarity of some early Proterozoic associations with modern "geosynclinal" assemblages has resulted in strongly divergent interpretations where one opinion holds that the full modern Wilson-cycle was in operation, while the other opinion favors "ensialic" tectonics with comparatively little relative motion of crustal segments now bordering these mobile belts (see Rutland, 1982 and Medaris, 1983 for full discussion). These contrasting points of view have now largely been reconciled by the general acceptance of considerable sialic underplating (A-subduction) during orogeny as presently documented in the Himalayas, and by the recognition of plate tectonicians that several Phanerozoic orogenic belts such as the Alps and Pyrenees appear to have resulted from largely intracontinental deformation whereby severely attenuated continental lithosphere was shortened and thickened by a variety of mechanisms ranging from thrust restacking to transform shearing (Dewey, 1982).

Le Pichon et al. (1982) and Dewey (1982) have discussed basin formation, lithospheric stretching and subsequent crust restacking while Kröner (1977, 1980, 1981b, 1983a, b) applied similar concepts in models for the evolution of African Proterozoic belts. None of these scenarios require a basin floored by oceanic crust, yet they are equally compatible with the plate tectonic concept as is the Wilson-cycle.

Following the speculation of Bird (1978) that subcrustal mantle lithosphere may delaminate during orogeny, Kröner (1983a) has developed a model for the ensialic evolution of Proterozoic mobile belts that includes the concepts discussed above and that infers subcrustal mantle rather than ocean crust subduction during horizontal shortening of previously attenuated continental crust. This model is presented in Figure 7 in a simplified form. It invokes lifting, heating and stretching of the crust as a result of lithospheric thinning over a mantle plume. This mechanism eventually leads to a "geosynclinal" basin floored entirely by continental crust. The rise of asthenosphere enhances gravitational instabilities in the old and dense subcrustal lithosphere and, on fracturing following crustal stretcing, results in spontaneous delamination. Hot asthenospheric material rises to take the place of the detached and sinking lithospheric slab, thereby inducing A-subduction and interstacking of previously attenuated continental crust. The much thickened crust is partially melted at depth, intruded by syn- and post-orogenic granites and finally uplifted and eroded to its present level of exposure. Episodic thermal anomalies during orogeny are caused by the rise of asthenospheric magmas to the base of the crust and by radioactive self-heating after crustal interstacking.

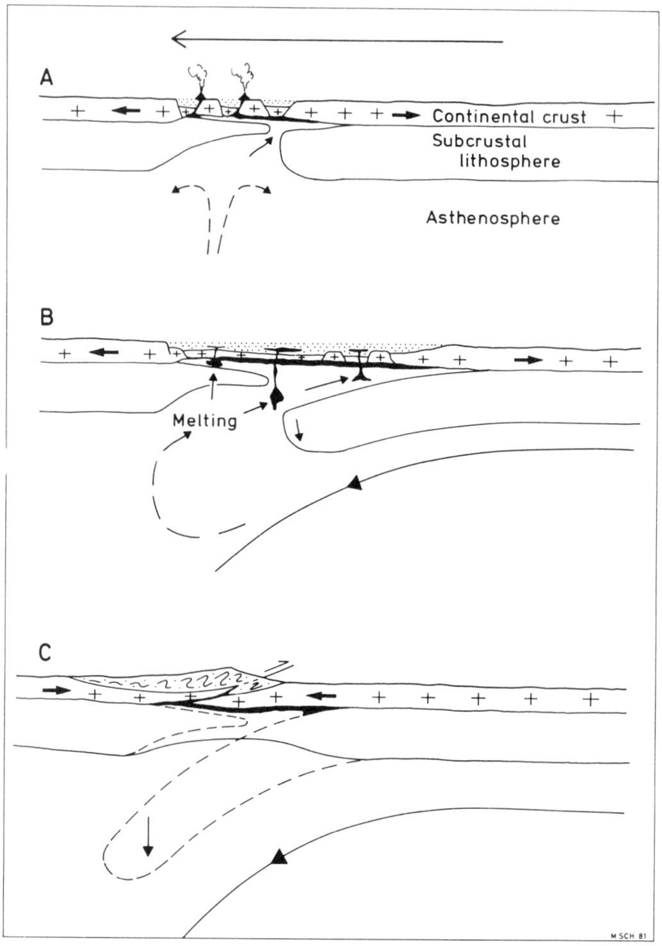

Fig. 7: Cartoon depicting three stages in the evolution of an "ensialic" foldbelt through lithospheric rifting and stretching, followed by delamination of subcrustal mantle lithosphere and crustal interstacking or A-subduction with horizontal shortening and orogeny. (From Kröner, 1983a).

The model is entirely compatible with the concept of horizontally moving plates but differs from the Wilson cycle in that no wet oceanic crust is generated during basin formation and none is consumed during orogeny. Instead, dry subcrustal lithosphere descends into the mantle but does not cause calc-alkaline magmatism, in line with the notable lack of such rocks in Proterozoic mobile belts. Bird (1978), Reutter et al. (1980), Fleitout and Froidevaux (1982) and Coman-Sad (1982) have proposed subcrustal lithospheric delamination for zones of Phanerozoic continental collision, and Solomon and

Head (1982) speculated that delamination may be a process occurring in convergence zones on Venus.

It is suggested that the majority of early Proterozoic mobile belts, particularly those originating between 2.1 and 1.8 Ga ago, evolved along lines as suggested above, i.e. without significant relative motion of cratonic blocks, and this is in line with the paleomagmetic record (for reviews see Kröner, 1981b; Rutland, 1982).

The concept of subcrustal delamination during orogeny implies that a substantial portion of old and dense lithosphere is removed, to be replaced by new asthenospheric and probably geochemcally fertile material. Cooling and thickening of this new layer to become rigid and stable lithosphere may require long equilibration periods as suggested by Pollack (1982), and the lithosphere under ensialic Proterozoic foldbelts may remain tectonically weak for several hundred million years. These linear zones of weakness are particularly susceptible for reactivation during release of intraplate stresses or they are the sites of future continental breakup.

In summary, continental growth during the lower Proterozoic was much less voluminous than in the Archaean and took place through some volcanic overplating and local horizontal accretion. The amount of subcrustal underplating is unknown, but if the Proterozoic rift-volcanics evolved in a similar manner to that postulated for Archaean greenstone belts, this process must have been important.

8. LATE PROTEROZOIC TECTONICS AND HORIZONTAL ACCRETION

The widespread recognition of late Proterozoic rock assemblages characterizing both passive and active continental margins as well as island-arc environments, together with the identification of true ophiolites and high-pressure mineral facies strongly suggests operation of modern-type Wilson-cycle processes towards the end of the Precambrian (for references see Kröner, 1980, 1981a). It remains one of the unresolved problems in crustal evolution, however, why these characteristic collision and accretion signatures are suddenly becoming so frequent in the geologic record.

Baer (1977) has suggested that the transition to the contemporary plate-tectonic regime was caused by the onset of subduction of oceanic crust, but this explanation remains unsatisfactory because the independent motion of continental plates since the Archaean implies that oceanic crust has been generated and subducted for at least the last 3 Ga. It appears more likely that the beginning of the Wilson-cycle is closely linked to individual plate velocities.

For example, the North American plate experienced its highest minimum velocity at around 1.7-1.8 Ga ago (Fig. 8a), at a time when an active continental margin evolved in the Wopmay orogen of the Slave Province (Hoffman, 1980). The African plate, however, moved at low minimum velocities during the early and middle Proterozoic (Fig. 8b) and experienced intraplate tectonics. Towards the end of the Proterozoic its velocity increased significantly and modern plate tectonics began.

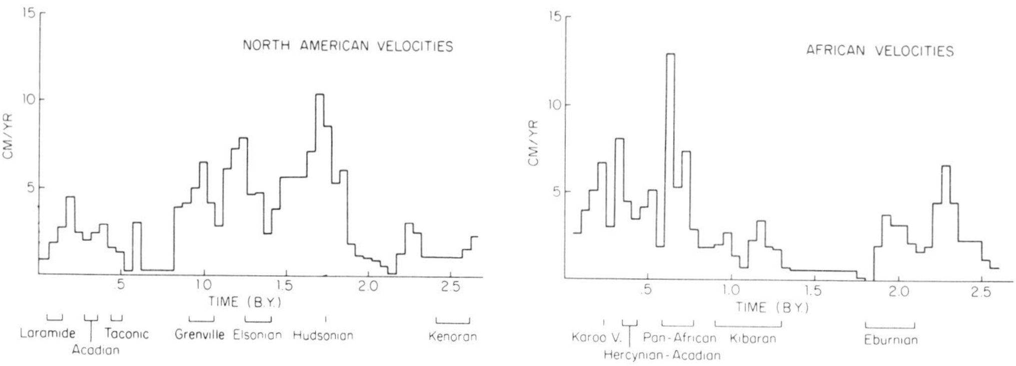

Fig. 8: Latitudinal plate velocities for North America (left) and for Africa (right) since the end of the Archaean. All velocities are given with respect to the pole. Note peak velocities at times of major orogenies (from Ullrich and Van der Voo, 1981).

In Africa continental accretion and ocean closure is documented by several suture-bounded ophiolites ranging in age between c. 900 Ma and about 650 Ma (Kröner, 1980). Particularly well documented areas are the margin of the West African Craton between Morocco and Mali (e.g. Caby et al., 1981) and the Arabian-Nubian shield of Egypt, Arabia and the Sudan (e.g. Al-Shanti, 1983).

Horizontal continental accretion in these terrains is revealed by juvenile island-arc evolution and thrust tectonics as a result of ophiolite obduction and arc collisions, but the most convincing evidence for lateral accretion of juvenile crust comes from isotope data. All isotopic systems support the contention that the island arc complexes of Arabia evolved essentially ensimatically from depleted but surprisingly homogeneous mantle (e.g. Fleck et al., 1980; Duyverman et al., 1982) and that further crustal differentiation through intra-crustal melting only occurred after arc collision and subduction-related lithospheric growth.

The late Proterozoic accretion model is now receiving an added complexity in that it is likely that exotic terrains that played a decisive role in the history of Phanerozoic Cordilleran (Coney et al., 1980) and Appalachian-type belts

(Williams and Hatcher, 1982) also existed in Proterozoic times. This was already suspected from paleomagnetic data in the Canadian shield (Irving and McGlynn, 1981) and is now also seriously discussed for the arc complexes of Arabia and northeast-Africa (Kröner et al., 1983 and unpubl. data).

Not all late Proterozoic foldbelts, however, fit the pattern of Wilson-cycle evolution (Wells et al., 1970; Kröner, 1977, 1980), and many are better explained by the model of ensialic orogeny and A-subduction as detailed above. For example, the Damara, Katanga and West Congo belts of Africa all contain thick miogeosynclinal and eugeosynclinal sediments but no ophiolites, calc-alkaline volcanics and blueschists (Kröner, 1983a). The difference between these intracontinental Upper Proterozoic belts and the older Lower Proterozoic mobile zones is evident in the fact that the former contain well developed thick geosynclinal sequences, signifying the formation of mature passive continental margins within wide basins bordering stable cratons, and sometimes even indicate limited crustal rupture with the emplacement of primitive tholeiitic lavas and mafic to ultramafic complex (Kröner, 1980; Strong, 1979). Also, the intensity of deformation, basement reactivation and thrust and nappe tectonics indicate greater horizontal shortening than in Lower Proterozoic belts and all these features are in accord with greater lithospheric mobility during Upper Proterozoic times.

Burke et al. (1977) have argued that all Precambrian foldbelts are due to plate collision mechanisms and that their apparently ensialic nature is due to considerable erosion that removed the diagnostic collision assemblages. However, paleomagnetic data do not support this contention (McWilliams, 1981), and at least for the Damara belt of Namibia Nd and Sr isotopes provide strong evidence for extensive reworking of older continental crust during the orogeny (Hawkesworth et al., 1981; Kröner, 1982).

In summary, evidence for both intracontinental orogeny and plate margin crustal accretion during the latest Precambrian support the contention that the period from about 900 Ma to about 550 Ma was one of global transition from predominantly intraplate to predominantly plate margin tectonics and crustal growth. It is possible that this transition to the present plate tectonic regime was strongly influenced by a major plate rearrangement that involved significant changes in the direction and velocity of plate motion.

9. DISCUSSION AND CONCLUSIONS

Massive and rapid lithospheric growth in the early Precambrian is documented by isotopic data and by geological evidence indicating the existence of at least a few continental segments that reached dimensions of several hundred to more than 1000 km (Kröner, 1981c) by the end of the Archaean, comparable in size to the dimensions of presently exposed Archaean cratons and to small continents that were apparently widespread in the Phanerozoic (Jones et al., 1982). If continental growth occurred through horizontal accretion in magmatic arcs, a distinct linearity and age zonation should be found in Archaean accretion belts. This, however, has not been demonstrated. Instead, almost all Archaean terrains indicate non-linear emplacement of granitoid rocks and, in some cases, a cyclic evolution pattern is evident in the magmatic history of granitoid terrains (Anhaeusser and Robb, 1981) that indicates secular changes in composition without any apparent geographic zonation. Also, the magmatic history in greenstone-granite terrains may be episodic and long, up to 800 Ma or more (e.g. Kaapvaal craton, Barton, 1981; Rhodesian craton, Nisbet et al., 1981).

All these features support the contention that long-lived thermal anomalies of considerable lateral extension must have existed below the early continents, and it is suggested that massive magmatic underplating was the cause for these anomalies and also provided the parent material for bimodal greenstone belt volcanism, as well as the tonalite-trondhjemite gneiss and granitoid association.

It is further speculated that extensive underplating was the prime cause for early and rapid lithospheric thickening that allowed large intracontinental basins to exist by 3.55 Ga ago. Perhaps the surprisingly low Archaean minimum average velocity of continental plates as postulated earlier is, to some extent, a reflection of considerable plate thickness and a lack of subduction-induced pull.

Time-integrated Sm/Nd ratios for volcanic rocks of the earliest known greenstone belts (3.8-3.5 Ga) suggest that these rocks were probably derived from a mantle that had already been depleted by even earlier major crust-forming events (Zindler, 1982; Jahn et al., 1982). If such crust formed in modern-style accretionary arcs, andesites should abound whereas, in fact, they are virtually non-existent in the oldest supracrustal suites.

It is suggested, therefore, that even the earliest continental crust did not originate and grow through intraoceanic arc evolution but primarily through hot-spot induced magmatism and iceland-type vertical accretion and differentiation (Kröner, 1981a, 1983b) that, at least locally, caused mantle depletion and early subcrustal lithospheric stabilization.

An early Archaean global heat flow of 200-250 mW/m^2 implies volcanic activity over virtually the entire surface of the early earth (Smith, 1981), and it is unlikely that, during the period of earliest lithosphere formation, this volcanism took place along linear zones comparable to present oceanic ridge systems. Rather, the closely spaced convection cells in a vigorously convecting mantle would suggest a system of thermal plumes (Kröner, 1981a; Lambert, 1981; Smith, 1981) or "hot lines" (Hanson, 1981).

Once volcanism produced sufficiently differentiated continental crust and a depleted "root zone" below, this early lithospheric segment could survive since its positive buoyancy overcame the drag forces of mantle convection (Hargraves, 1981). Regions of the presumably dense and largely komatiitic ocean crust that were not situated over plumes and therefore did not grow through over- and underplating were easily dragged down into the mantle since they became negatively buoyant within several tens of millions of years after their formation (Arndt, pers. comm., 1982).

Ocean crust recycling through subducton must have occurred since the earliest Archaean as demanded by Sm-Nd systematics (e.g. McCulloch and Compston, 1982), and the available evidence for independent plate motion since at least 3.5 Ga ago also supports this contention. However, in the early earth subducton was apparently not an effective crust accretion mechanism, perhaps because the mantle was hotter and subducted slabs may never have reached the depths to trigger extensive melting in the mantle wedge above. Modern island-arc volcanism probably requires subduction of oceanic lithosphere to a depth of at least 100 km (Solomon et al., 1981), a condition that may not have been possible during the Archaean.

It would appear from the available global isotopic record that the period between 2.8 Ga and 2.5 Ga ago constituted the most important crust-forming episode in earth's history, where massive underplating has created extensive reservoirs for granite-greenstone magmatism that extended well into the lower Proterozoic. It has been suggested that this event was related to a dramatic climax in the thermal evolution of the earth (Lambert, 1981) that was followed by substantial cooling and a reorganization in the mantle convection pattern (Fyfe, 1978).

Proterozoic crustal tectonics created large intracratonic basins as well as mobile belts. The former can be explained by thermally induced abortive attempts of the thickened lithosphere to break apart during which magmatic underplating again provided the reservoirs for intracontinental, largely bimodal volcanism. The latter may have evolved on thinner lithosphere and not previously stabilized crust that responded to linear thermal anomalies from below through considerable stretching, thinning and "geosynclinal" basin formation.

It is remarkable that many Proterozoic mobile belts lack a distinctive magmatic arc accretionary rock assemblage, and this implies that plate separation and subsequent subduction of significant ocean crust during basin closure did not occur, at least not at the scale required for arc magmatism. Most probably these belts developed from intracontinental or small ocean basins of Red-Sea type through crust restacking and A-subduction. Only in rare cases did active continental margins form, and here the first significant amounts of continental crust were laterally accreted as, for example, the Wopmay orogen of Canada and the Svecofennian of Scandinavia.

The model of subcrustal delamination as advocated earlier provides the thermal conditions for lower crustal heating and extensive intracrustal melting that characterize many Proterozoic terrains, and also permits the formation of new "juvenile" lithosphere through underplating. Houseman et al. (1981) have suggested an alternative mechanism to trigger delamination of subcrustal lithosphere. In their model delamination occurs during the stage of intracontinental crustal shortening where crustal thickening is accompanied by an equal amount of subcrustal lithospheric thickening. The relatively cold lower lithosphere is therefore pushed downwards into warmer asthenosphere, thereby enhancing the gravitational instability between these two layers. This may eventually lead to rapid detachment and sinking of the thickened cold mantle layer with concomitant rise of asthenospheric material and subsequent heating of the lower crust now stripped from its insulating base.

These arguments may show that Proterozoic tectonics differed markedly from that in earlier times, but that lithospheric growth continued to occur predominantly through underplating, in some cases perhaps by the replacement of old, dense subcrustal lithosphere through juvenile and "fertile" material that may have become the source for later intracontinental magmatic assemblages.

Towards the end of the Precambrian, crustal growth by lateral accretion began to dominate lithospheric evolution and followed a mechanism that has remained essentially unchanged for the last 900 Ma. However, there is no evidence that all growth since that time was solely by horizontal accretion, and models for the evolution of several Phanerozoic orogenic belts such as the Pyrenees, the Hercynides and the Alps are more in line with the closure of small basins and limited or no ocean crust subduction than with operation of the full Wilson-cycle (Kröner, 1983a; Dewey, 1982). In these cases as well as during the initial stages of Gondwana continental breakup, significant underplating by vertical accretion must have occurred while elsewhere horizontal growth took place at the same time.

It is concluded that vertical rather than horizontal growth of lithosphere has dominated global evolution during the first 3.5 Ga of earth's history and that this mechanism also accounts for the observed magmatic rock assemblages as well as for lithospheric stabilization and cratonization of continental crust. Crustal differentiation does not necessarily require horizontal accretion as also shown by the suggested evolution of one-plate planets such as Venus and Mars (Solomon et al., 1981)

ACKNOWLEDGMENTS

The West German Research Council (DFG) has funded most of the resarch and travel that led to the views expressed in this paper, and I also acknowledge the stimulating discussions with many colleagues during the activities of IGCP Project 92 (Archaean Geochemistry).

This review is a contribution to the DFG-supported Research Group "Accretion and Differentiation of the Earth" at Mainz and to the Lithosphere Program. I thank A. W. Hofmann for constructive cristicism of the manuscript.

REFERENCES

Al-Shanti, A. A. (ed.), (1983) Pan-African crustal evolution in Arabia and north-east Africa. Fac. Earth Sci., King Abdulaziz Univ., Jeddah, Bull. 6 (in press).

Anhaeusser, C. R. (1971) The Barberton Mountain land, South Africa - a guide to the understanding of the Archaean geology of western Australia. Geol. Soc. Australia, Spec. Publ. 3, 103-119.

Anhaeusser, C. R. (1981) Geotectonic evolution of the Archaean successions in the Barberton Mountain Land, South Africa. In: Kröner, A. (ed.) Precambrian plate tectonics. Elsevier, Amsterdam, 137-160.

Archibald, N.J., Bettenay, L. F., Bickle, M. J. and Groves, D. I. (1981) Evolution of Archaean crust in the eastern Goldfields Province of the Yilgarn block, Western Australia. In: Glover, J. E. and Groves, D. I. (eds.) Archaean Geology. Geol. Soc. Australia Spec. Publ., 7, 491-504.

Arndt, N. T. and Nisbet, E. H. (eds.) (1982) Komatiites. George Allen & Unwin, London, 526 p.

Ashwal, L. D., Phinney, W. C. and Wooden, J. L. (1982) Underplating of Archaean continents: Evidence from the Bad Vermilion Lake anorthosite complex, Ontario. Lunar Planet. Sci., 13, Lunar and Planetary Institute, Houston, 20-21.

Baer, A. J. (1977) Speculations on the evolution of the lithosphere. Precambrian Res., 5, 249-260.

Baer, A. J. (1981) Geotherms, evolution of the lithosphere and plate tectonics. Tectonophysics, 72, 203-227.

Baragar, W. R. A. (1966) Geochemistry of the Yellowknife volcanic rocks. Can. J. Earth Sci., 3, 9-30.

Barker, F. and Arth, J. G. (1976) Generation of trondhjemitic-tonalitic liquids and Archaean bimodal tranodhjemite-basalt suites. Geology, 4, 596-600.

Barker, F., Arth, J. G. and Hudson, T. (1981) Tonalites in crustal evolution. Phil. Trans. R. Soc. Lond., A301, 185-187.

Barton, Jr., J. M. (1981) The pattern of Archaean crustal evolution in southern Africa as deduced from the evolution of the Limpopo mobile belt and the Barberton granite-greenstone terrain. In: Glover, J. E. and Groves, D. I. (ed.) Archaean geology. Geol. Soc. Australia, Spec. Publ. 7, 21-32.

Bettenay, L. F., Bickle, M. J., Boulter, C. A., Groves, D. I., Morant, P., Blake, T. S. and James, B. A. (1981) Evolution of the Shaw batholith - an Archaean granitoid-gneiss dome in the eastern Pilbara, Western Australia. In: Glover, J. E. and Groves, D. I. (eds.) Archaean Geology. Geol. Soc. Australia Spec. Publ. 7, 361-372.

Bickle, M. J. (1978) Heat loss from the earth: constraints on Archaean tectonics from the relation between geothermal gradients and the rate of plate production. Earth Planet. Sci. Lett., 40, 301-315.

Bickle, M. J. and Eriksson, K. A. (1982) Evolution and subsidence of early Precambrian sedimentary basins. Phil. Trans. R. Soc. Lond., A305, 225-247.

Bird, P. (1978) Initiation of intracontinental subduction in the Himalaya. J. Geophys. Res., 83, 4975-4987.

Burke, K., Dewey, J. F. and Kidd, W. S. F. (1977) World distribution of sutures - the sites of former oceans. Tectonophysics, 40, 69-99.

Caby, R., Bertrand, J. M. L. and Black, R. (1981) Pan-African ocean closure and continental collision in the Hoggar-Iforas segment, central Sahara. In: Kroner, A. (ed.), Precambrian plate tectonics. Elsevier, Amsterdam, 407-434.

Colman-Sadd, S. P. (1982) Two stage continental collision and plate driving forces. Tectonophysics, 90, 263-282.

Condie, K. C. (1975) A mantle plume model for the origin of Archaean greenstone belts based on trace element distributions. Nature, 258, 413-414.

Condie, K. C. (1976) Trace element geochemistry of Archaean greenstone belts. Earth Sci. Rev., 12, 393-417.

Condie, K. C. (1980) Origin and early development in the earth's crust. Precambrian Res., 11, 183-197.

Condie, K. C. (1981a) Archaean greenstone belts. Elsevier, Amsterdam, 434p.

Condie, K. C. (1981b) Geochemical and isotopic constraints on the origin and source of Archaean granites. In: Glover, J. E. and Groves, D. I. (eds.) Archaean geology. Geol. Soc. Australia, Spec. Publ. 7, 469-480.

Condie, K. C. (1982) Early and middle Proterozoic supracrustal successions and their tectonic settings. Am. J. Sci., 282, 341-357.

Condie, K. C. and Hunter, D. R. (1976) Trace element geochemistry of Archaean granitic rocks from the Barberton region, south Africa. Earth Planet. Sci. Lett., 29, 289-400.

Coney, P. J., Jones, D. L. and Monger, J. W. H. (1980) Cordilleran suspect terranes. Nature, 288, 329-333.

Cook, F. A. and Turcotte, D. L. (1981) Parameterized convection and the thermal evolution of the earth. Tectonophysics, 75, 1-17.

Cordani, U. and Yier, S. S. (1978) Geochronological investigation on the Precambrian granulite terrains of Bahia, Brazil. Precambrian Res., 9, 255-274.

Cox, K. G. (1980) A model for flood basalt volcanism. Jour. Petrol., 21, 629-650.

Davies, G. F. (1979) Thickness and thermal history of continental crust and root zones. Earth Planet. Sci. Lett., 44, 231-238.

Davies, G. F. and Strebeck, J. W. (1982) Old continental geotherms: constraints on heat production and thickness of continental plates. Geophys. J. R. astr. Soc., 69, 623-634.

De Laeter, J. R., Libby, W. G. and Trendall, A. F. (1981) The older Precambrian geochronology of Western Australia. In: Glover, J. E. and Groves, D. I. (eds.), Archaean geology. Geol. Soc. Australia, Spec. Publ. 7, 145-157.

DePaolo, D. J. (1981) Neodymium isotopes in the Colorado Front Range and crust-mantle evolution in the Proterozoic. Nature, 291, 193-196.

DePaolo, D. J. (1982) Disturbed Sm-Nd and Rb-Sr systems in altered late Archaean basalts of the Fortescue Group, Western Australia. Geol. Soc. America, Abstr. with Progr., 14, 473.
Dewey, J. F. (1982) Plate tectonics and the evolution of the British Isles. J. Geol. Soc. London, 139, 371-412.
Dewey, J. F. and Windley, B. F. (1981) Growth and differentiation of the continental crust. Phil. Trans. R. Soc. Lond. A301, 189-206.
De Wit, M. J. (1982) Gliding and overthrust nappe tectonics in the Barberton greenstone belt. Jour. Struct. Geol., 4, 117-136.
Dickinson, W. R. (1981) Plate tectonics through geologic time. Phil. Trans. R. Soc. Lond. A301, 207-215.
Dunlop, D. J. (1981) Paleomagnetic evidence for Proterozoic continental development. Phil. Trans. R. Soc. Lond. A301, 265-277.
Duyverman, H. J., Harris, N. B. W. and Hawkesworth, C. J. (1982) Crustal accretion in the Pan African: Nd and Sr isotope evidence from the Arabian shield. Earth Planet. Sci. Lett., 59, 315-326.
East Pacific Rise Study Group (1981) Crustal processes of the mid-ocean ridge. Science, 213, 31-40.
Eriksson, K. A. (1981) Archaean platform-to-trough sedimentation, east Pilbara Block, Australia. In: Glover, J. E. and Groves, D. I. (eds.) Archaean Geology. Geol. Soc. Australia, Spec. Publ. 7, 235-244.
Ewart, A., Baxter, K. and Ross, J. A. (1980) The petrology and petrogenesis of the Tertiary anorgenic mafic lavas of southern and central Queensland, Australia- possible implications for crustal thickening. Contrib. Mineral. Petrol., 75, 129-152.
Fleck, R. J., Greenwood, W. R., Hadley, D. G., Anderson, R. E. and Schmidt, D. L. (1980) Rubidium-strontium geochronology and plate-tectonic evolution of the southern part of the Arabian shield. U.S. Geol. Surv. Prof. Paper 1131, 38 p.
Fleitout, L. and Froidevaux, C. (1982) Tectonics and topography for a lithosphere containing density heterogeneities. Tectonics, 1, 21-56.
Forsyth, D. and Uyeda, S. (1975) On the relative importance of driving forces of plate motions. Geophys. J. R. astron. Soc., 43, 103-200.
Fripp, R. E. P. (1981) The ancient Sand River Gneisses, Limpopo mobile belt, South Africa. In: Glover, J. E. and Groves, D. I. (eds.) Archaean Geology. Geol. Soc. Australia, Spec. Publ. 7, 329-336.
Fyfe, W. S. (1978) The evolution of the earth's crust: modern plate tectonics to ancient hot spot tectonics? Chem. Geol., 23, 89-114.
Gastil, R. G. (1979) A conceptual hypothesis for the relation of differing tectonic terranes to plutonic emplacement. Geology, 7, 542-544.
Geary, E. E. and Kay, R. W. (1982) Petrologic and geochemical documentation of ocean floor metamorphism in the Zambales ophiolite, Philippines. Am. Geophys. Union, Geophys. Monogr. Ser., 27.
Gee, R. D., Baxter, J. L., Wilde, S. A. and Williams, I. R. (1981) Crustal development in the Archaean Yilgarn Block, western Australia. In: Glover, J. E. and Groves, D. I. (eds.) Archaean Geology. Geol. Soc. Austra., Spec. Publ. 7, 43-56.
Giles, C. W. and Hallberg, J. A. (1982) The genesis of the Archaean Welcome Well volcanic complex, Western Australia. Contrib. Mineral. Petrol., 80, 307-318.
Gill, J. (1981) Orogenic andesites and plate tectonics. Springer-Verlag, Berlin, 390p.
Glikson, A. Y. (1976) Earliest Precambrian ultramafic -mafic volcanic rocks: ancient oceanic crust or relict terrestrial maria? Geology, 4, 201-206.
Glikson, A. Y. (1979) Early Precambrian tonalite-trondhjemite sialic nuclei. Earth Sci. Rev., 15, 1-73.
Glover, J. E. and Groves, D. I. (eds.) (1981) Archaean Geology, Geol. Soc.

Australia, Spec. Publ. 7, 515p.
Goodwin, A. M. (1981) Archaean plates and greenstone belts. In: Kröner, A. (ed.) Precambrian plate tectonics. Elsevier, Amsterdam, 105-135.
Grachev, A. F. and Fedorovsky, V. S. (1981) On the nature of greenstone belts in the Preambrian. Tectonophysics, 73, 195-212.
Green, D. H. (1975) Genesis of Archaean peridotitic magmas and constraints on Archaean geothermal gradients and tectonics. Geology, 3, 15-18.
Green, D. H. (1981) Petrogenesis of Archaean ultramafic magmas and implications for Archaean tectonics. In: Kröner, A. (ed.) Precambrian plate tectonics. Elsevier, Amsterdam, 469-489.
Greenwood, W. R., Stoeser, D. B., Fleck, R. J. and Stacey, J. S. (1982) Late Proterozoic island arc complexes and tectonic belts in the southern part of the Arabian shield. U. S. Geol. Surv., Saudi Arabia Mission, Open file Rept. USGS OF-02-8, 46 pp.
Hager, B. H. and O'Connell, R. J. (1980) Lithosphere thickening and subduction, plate motions and mantle convection. Physics of the earth's interior, course 78. Soc. Italiana di Fisica. North-holland, Amsterdam, 464-492.
Hallbauer, D. K., Kable, E. and Robinson, D. N. (1980) The occurrence of detrital diamond and moissanite in the Proterozoic Witwatersrand conglomerates and their implication for crustal evolution. Abstr.-vol., Internat. Wegener Symposium, Berlin. Berlin. Geowiss. Abh. A19, 67-68.
Hamilton, P. J., Evensen, N. M. and O'Nions, R. K. (1979) Sm-Nd systematics of Lewisian gneisses: implciations for the origin of granulites. Nature, 277, 25-28.
Hamilton, W. (1981) Crustal evolution by arc magmatism. Phil. Trans. R. soc. Lond., A301, 279-291.
Hanson, G. N. (1981) Geochemical constraints on the evolution of the early continental crust. Phil. Trans. R. Soc. Lond., A301, 423-442.
Hargraves, R. B. (1981) Precambrian tectonic style: a liberal uniformitarian interpretaton. In: Kröner, A. (ed.) Precambrian plate tectonics. Elsevier. Amsterdam, 21-56.
Hart, R. J., Welke, H. J. and Nicolaysen, L. O. (1981) Geochronology of the deep profile through Archaean basement at Vredefort, with implications for early crustal evolution. Jour. Geophys. Res., 86, 10663-10680.
Hart, S. R., Brooks, C., Krogh, T. E., Davis, G. L. and Nava, D. (1970). Ancient and modern volcanic rocks: a trace element model. Earth Planet. Sci. Lett., 10, 17-28.
Hawkeswroth, C. J. and O'Nions, R. K. (1977) The petrogenesis of some Archaean volcanic rocks from southern Africa. J. Petrol., 18, 487-520.
Hawkesworth, C. J., Kramers, J. D. and Miller, R. McG. (1981). Old model Nd ages in Namibian Pan-African rocks. Nature, 289, 278-281.
Hegner, E., Tegtmeyer, A. and Kröner, A. (1981) Geochemie and Petrogenese archaischer Vulkanite der Pongola Gruppe in Natal, Sudafrika. Chem. Erde, 40, 23-57.
Heier, K. S. and Thorensen, K. (1971) Geochemistry of high grade metamorphic rocks, Lofoten-Vesteralen, North Norway. Geochim. Cosmochim. Acta, 35, 89-99.
Hekinian, R. (1974) Petrology of igneous rocks from leg 22 in the north-eastern Indian ocean. Initial Repts. of the Deep Sea Drilling Project, 22. U.S. Govt. Printing Office, Washington, D. C., 413-447.
Helz, R. T. (1976) Phase relations of basalts in their melting ranges at P_{H_2O} = 5kb. Part II. Melt compositions. Jour. Petrol., 17, 139-193.
Hickman, A. H. (1981) Crustal evolution of the Pilbara Block, western Australia. In: Glover, J. E. and Groves, D. E. (eds.) Archaean Geology. Geol. Soc. Australia, Spec. Publ. 7, 57-69.
Hietanen, A. (1975) Generation of potassium-poor magmas in the northern Sierra Nevada and the Svecofennian of Finland. Jour. Res. U.S. Geol. Surv., 3, 631-645.

Hoffman, P. (1980) Wopmay orogen: a Wilson cycle of early Proterozoic age in northwest of Canadian shield. Geol. Ass. Canada, Spec. Paper 20, 523-549.
Hofmann, A. W. and White, W. M. (1982) Mantle plumes from ancient oceanic crust. Earth Planet. Sci. Lett., 57, 421-436.
Horwitz, R. C. and Smith, R. E. (1978) Bridging the Yilgarn and Pilbara Blocks, Western Australia. Precambrian Res., 6, 293-322.
Housman, G. A., McKenzie, D. P. and Molnar, P. (1981) convective instability of a thickened boundary layer and its relevance for the thermal evolution of continental convergence belts. J. Geophys. Res., 86, 6115-6132.
Hurley, P. M. and Rand, J. R. (1969) Pre-drift continental nucleii. Science, 164, 1229-1242.
Hynes, A. and Francis, D. M. (1982) A transect of the early Proterozoic Cape Smith foldbelt, New Quebec. Tectonophysics, 88, 23-59.
Irving, E. and McGlynn, I. C. (1981) On the coherence, rotation and paleolatitude of Laurentia in the Proterozoic. In: Kröner, A. (ed.) Precambiran plate tectonics. Elsevier, Amsterdam, 561-598.
Jackson, M. P. A. and Robertson, D. I. (1983) Regional implications of early-Precambrian strains on the Onverwacht Group adjacent to the Lochiel Granite, north-west Swaziland. In: Anhaeusser, C. R. (ed.), Contributions to the geology of the Barberton Mountain Land. Geol. Soc. S. Afr. Spec. Publ. 9, 45-62.
Jacobsen, S. B. and Wasserburg, G. J. (1979) The mean age of mantle and crustal reservoirs. J. Geophys. Res., 84. 7411-7427.
Jahn, B.-M., Shih, C. Y. and Murthy, V. R. (1974). Trace element geochemistry of Archean volcanic rocks. Geochim. Cosmochim. acta, 38, 611-627.
Jahn, B.-M., Grauau, G. and Glikson, A. Y. (1982). Komatiites of the Onverwacht Group, S. Africa: REE geochemistry, Sm/Nd age and mantle evolution. Contrib. Mienral. Petrol., 80, 25-40.
Jahn, B.-M. and Sun, S. S. (1979) Trace element distribution and isotopic composition of Archean greenstones. In: Ahrens, L. H. (ed.) Origin and distribution of the elements. Pergamon Press, Oxford, 597-618.
Jahn, B.-M., Vidal, P. and Kroner, A. (1984). Sm-Nd/Pb-Pb ages and REE geochemistry of the Tojottamanselka gneisses of northern Finland. Contrib. Mineral. Petrol., in press.
Jones, D. L., Cox, A., Coney, P. and Beck, M. (1982). The growth of western North America. Scientific American, 247, 70-84.
Jordan, T. H. (1979) The deep structure of continents. Scientific American, 240, 70-82.
Jordan, T. H. (1981) Continents as a chemical boundary layer. Phil. Trans. R. Soc. Lond., A301, 359-373.
Katz, M. B. (1977) Early Precambrian granulites-greenstones, transform mobile belts and their tectonic setting. In: Sidorenko, A. V. (ed.) correlation of the Precambrian, 1. Nauka Publ. Office, Moscow, 26-39.
Kramers, J. D. (1979) Lead uranium, strontium, potassium and rubidium in inclusion-bearing diamonds and mantle-derived xenoliths from southern Africa. Earth Planet. Sci. Lett., 42, 58-70.
Kröner, A. (1977) Precambrian mobile belts of southern and eastern Africa - ancient sutures or sites of ensialic mobility? A case for crustal evolution towards plate tectonics. Tectonophysics, 40, 101-135.
Kröner, A. (1980) Pan-African crustal evolution. EPISODES, 1980, 2, 3-8.
Kröner, A. (1981a) Precambrian plate tectonics. In: Kröner, A. (ed.) Precambrian plate tectonics. Elsevier, Amsterdam, 57-90.
Kröner, A. (ed.) (1981b) Precambrian plate tectonics. Elsevier, Amsterdam. 781 p.
Kröner, A. (1981c) Precambrian crustal evolution and continental drift. Geol. Rundschau, 70, 412-428.

Kröner, A. (1982) Rb-Sr geochronology and tectonic evolution of the Pan-African Damara belt of Namibia, southwestern Africa. Am. J. Sci., 282.
Kröner, A. (1983a) Proterozoic mobile belts compatible with the plate tectonic concept. In: Medaris, Jr., G. L. (ed.), Proterozoic geology: Selected papers from an International Proterozoic Symposium. Geol. Soc. America Memoir 161, 59-74.
Kröner, A. (1983b) Archaean to early Proterozoic tectonics and crustal evolution: a review. Rev. Brasil. Geocienc., 12, 15-31.
Kröner, A. and McWilliams, M. O. (1982) Archean paleomagnetism of the Kaapvaal craton, southern Africa, and implications for early Archean tectonics. Geol. Soc. America, Abstr. with programs, 14, 536.
Kröner, A., Stern, R. J. and Fleck, R. J. (1983) Chronology and mechanism of Pan-African crustal accretion in northeast Africa and Arabia. Abstr.-vol., 12th Colloquium Afr. Geol., Brussels, Belgium, Mus. royal Afr. centr., 55.
Lambert, R. St. J., (1981) Earth tectonics and thermal history: review and a hot-spot model for the Archaean. In: Kroner, A. (ed.) Precambrian plate tectonics. Elsevier, Amsterdam, 453-467.
Le Pichon, X., Angelier, J. and Sibuet, J. C. (1982) Plate boundaries and extensional tectonics. Tectonophysics, 81, 239-256.
McCulloch, M. T. and Wasserburg, G.J. (1978) Sm-Nd age of Kambalda and Kanowna greenstones and heterogeneity in the Archaean mantle. Nature, 294, 322-327.
McElhinny, M. W. and Senanayake, W. E. (1980) Paleomagnetic evidence for the existence of the geomagnetic field 3.5 Ga ago. J. Geophys. Res., 85, 3523-3528.
McGregor, V. R. (1979) Archaean grey gneisses and the orogin of the continental crust: evidence from the Godhaab region, West Greenland. In: Barker, F. (ed.) Trondhjemites, dacites and related rocks. Elsevier, Amsterdam, 169-204.
McIver, J. R., Cawthorn, R. G. and Wyatt, B.A. (1982) The Ventersdorp Supergroup - the youngest komatiitic sequence in South Africa. In: Arndt, N. T. and Nisbet, E. G. (eds.) Komatiites. George Allen & Unwin, Lond, 81-90.
McKenzie, D. (1978) Some remarks on the development of sedimentary basins. Earth Planet. Sci. Lett., 40, 25-32.
McLennan, S. M. and Taylor, S. R. (1982) Geochemical constraints on the growth of the continental crust. Jour. Geol., 90, 347-361.
McLennan, S. M., Taylor, S. R. and Kröner, A. (1983) Geochemical evolution of Archaean shales from southern Africa: The Swaziland and Pongola Supergroups. Precambrian Res., 22, 93-124.
McWilliams, M. O. (1981) Paleomagnetism and Precambrian tectonic evolution of Gondwana. In: Kröner, A. (ed.) Precambrian plate tectonics. Elsevier, Amsterdam, 649-687).
Medaris Jr., G. L. (ed.) (1983) Proterozoic geology: Selected papers from an International Proterozoic Symposium. Geol. Soc. America, Memoir 161.
Morgan, W. J. (1971) Convection plumes in the lower mantle. Nature, 230, 42-43.
Myers, J. S. (1976) Granitoid sheets, thrusting, and Archaean crustal thickening in West Greenland. Geology, 4, 265-268.
Moorbath, S. (1978) Age and isotope evidence for the evolution of the continental crust. Phil. Trans. R. Soc. Lond. A288, 401-413.
Newton, R. C., Smith, J. V. and Windley, B. F. (1980) Carbonic metamorphism, granulites and crustal growth. Nature, 288, 45-49.
Newton, R. C. and Hanson, E. C. (1983) The origin of Precambrian granulite terrains - evidence from the transition zones. In Medaris Jr., G. L. (ed.) Proterozoic geology: Selected papers from an International Proterozoic Symposium. Geol. Soc. America, Memoir 161.
Nisbet, E. G. (1982) The tectonic setting and petrogenesis of komatiites. In: Arndt, N. T. and Nisbet, E. G. (eds.) Komatiites. George Allen & Unwin, London, 501-520.

Nisbet, E. G., Wilson, J. F. and Bickle, M. J. (1981) The evolution of the Rhodesian craton and adjacent Archaean terrain; tectonic models. In: Kröner, A. (ed.) Precambrian plate tectonics. Elsevier, Amsterdam, 161-183.

Oskarsson, N., Sigvaldason, G. E. and Steinthorsson, S. (1982) A dynamic model of rift zone petrogenesis and the regional petrology of Iceland. Jour. Petrol., 23, 28-74.

Oxburgh, E. R. (1981) Heat flow and difference in lithospheric thickness. Phil. Trans. R. Soc. Lond., A 301, 337-346.

Padgham, W. A. (1981) Archaean crustal evolution - a glimpse from the Slave Province. In: Glover, J. E. and Groves, D. I. (eds.) Archaean Geology. Geol. soc. Australia, Spec. Publ. 7, 99-110.

Phinney, W. C., Morrison, D. A. and Ashwal, L. D. (1981) Implications of Archaean anorthosites for crust-mantle evolution. Lunar Planet. Sci., 12, Lunar and Planetary Institute, Houston, 830-832.

Piper, J. D. A. (1982) The Precambrian paleomagnetic record. The case for the Proterozoic supercontinent. Earth Planet. Sci. Lett., 59, 61-89.

Platt, J. P. (1980) Archaean greenstone belts: a structural test of tectonics hypotheses. Tectonophysics, 65, 127-150.

Pollack, H. N. (1982) The heat flow from the continents. Ann. Rev. Earth Planet. Sci., 1982, 10, 459-481.

Prestvik, T. (1982) Basic volcanic rocks and tectonic setting. A discussion of the Zr-Ti-Y discrimination diagram and its suitability for classification purposes. Lithos, 15, 241-247.

Reutter, K. J., Giese, P. and Closs, H. (1980) Lithospheric split in the descending plate: observations from the northern Apennines. Tectonophysics, 40, T1-T9.

Ringwood, A. E. (1982). Phase transformations and differentiation in subducted lithosphere: implication for mantle dynamics, basalt petrogenesis and crustal evolution. Jour. Geol., 90, 611-644.

Rutland, R. W. R. (1982) On the growth and evolution of the continental crust: a comparative tectonic approach. Jour. & Proc. R. Soc. N. S. W., 115, 33-60.

Scharer, U. and Allegre, C. J. (1982) Investigation of the Archaean crust by single-grain dating of detrital zircon: a graywacke of the Slave Province, Canada. Can. J. Earth Sci., 19, 1910-1918.

Sighinolfi, G. P., Figueiredo, M. C. H., Fyfe, W. S., Kronberg, B. I. and Oliveira, M. A. F. T. (1981) Geochemistry and petrology of the Jequie granulitic complex (Brazil): an Archaean granitic basement complex. Contrib. Min. Petro., 87, 263-271.

Sigurdsson, H. and Sparks, R. S. J. (1981) Petrology of rhyolitic and mixed magma ejecta from the 1875 eruption of Askja, Iceland. Jour. Petrol., 22, 41-84.

Sleep, N. H. and Windley, B. F. (1982) Archaean plate tectonics: constraints and inferences. Jour. Geol., 90, 363-380.

Smith, J. V. (1981) The first 800 million years of Earth's history. Phil. Trans. R. Soc. Lond. A301, 401-422.

Solomon, S. C., Ahrens, T. J., Cassen, P. M., Hsui, A. T., Minear, J. W., Reynolds, R. T., Sleep, N. H., Strangway, D. W. and Turcotte, D. L. (1981) Thermal histories of the terrestrial planets. In: Basaltic Volcanism Study Project (eds.) Basaltic volcanism on the terrestrial planets. Pergamon Press, New York, 1129-1234.

Solomon, S. C. and Head, J. W. (1982) Mechanisms for lithospheric heat transport on Venus: implications for tectonic style and volcanism. J. Geophys. Res., 87, 9236-9246.

Stowe, C. W. (1974) Alpine-type structures in the Rhodesian basement complex at Selukwe. J. Geol. Soc. London, 130, 411-425.

Strong, D. F. (1979) Proterozoic tectonism of northwestern Gondwanaland: new evidence from eastern Newfoundland. Tectonophysics, 54, 81-101.

Talwani, M. and Langseth, M. (1981) Oceanic crustal dynamics. Science, 213, 22-31.
Tankard, A. J., Jackson, M. P. A., Eriksson, K. A., Hobday, D. K., Hunter, D. R. and Minter, W. E.L. (1982) Crustal evolution of southern Africa - 3.5 billion years of earth history. Springer-Verlag, New York, 523 p.
Tarney, J., Dalziel, I. W. D. and de Wit, M. J. (1976) Marginal basin "Rocas Verdes" complex from S. Chile: a model for Archaean greenstone belt formation. In: Windley, F. B. (ed.) The early history of the earth. Wiley & Sons, London, 131-146.
Taylor, S. R. (1967) the origin and growth of continents. Tectonophysics, 4, 17-34.
Thompson, G., Bryan, W. B., Frey, F. A. and Dickey Jr., J. S. (1978) Basalts and related rocks from deep-sea drilling sites in the central and eastern Indian ocean. Marine Geol., 26, 119-138.
Thorpe, R. S. (ed.) (1982) Andesites: orogenic andesites and related rocks. John Wiley & Sons, Chichester, 724 p.
Ullrich, L. and Van der Voo, R. (1981) Minimum continental velocities with respect to the pole since the Archaean. Tectonophysics, 74, 17-27.
Viljoen, M. J. and Viljoen, R. P. (1969) An introduction to the geology of the Barberton granite-greenstone terrain. Geol. Soc. S. Afr., Spec. Publ. 2, 9-28.
Weaver, B. L. and Tarney, J. (1982) Andesitic magmatism and continental growth. In: Thorpe, R. S. (ed.) andesites: Orogenic andesites and related rocks. Wiley & Sons, Chichester, 639-661.
Wells, A. T., Forman, D. J., Ranford, L. C. and Cook, P. J. (1970) Geology of the Amadeus basin, central Australia. BMR Geol. Geophys., Bull. 100, 222 p.
Wells, P. R. A. (1981) Accretion and continental crust: thermal and geochemical consequences. Phil. Trans. R. Soc. Lond., A301, 347-357.
Wilkinson, J. F. G. and Binns, R. A. (1977) Relatively iron-rich lherzolite xenoliths of the Cr-diopside suite: a guide to the primary nature of anorogenic tholeiitic andesite magmas. Contrib. Mineral. Petrol., 65, 199-212.
Williams, H. and Hatcher Jr., R. D. (1982) Suspect terranes and accretionary history of the Appalachian orogen. Geology, 10, 530-536.
Wilson, J. F., Bickle, M. J., Hawkesworth, C. J., Martin, A., Nisbet, E. G. and Orpen, J. L. (1978) Granite-greenstone terrain of the Rhodesian Archaean craton. Nature, 271, 23-27.
Wilson, M. R. (1982) Magma types and the tectonic evolution of the Swedish Proterozoic. Geol. Rundschau, 71, 120-129.
Windley, B. F. (ed.) (1976) The early history of the earth. Wiley, London, 619 p.
Windley, B. F. (1977) The evolving continents. Wiley & Sons, London, 385 p.
Windley, B. F. (1979) Tectonic evolution of continents in the Precambrian. Episodes, 1979, 4, 12-16.
Windley, B. F. (1981) Precambrian rocks in the light of the plate tectonic concept. In: Kröner, A. (ed.) Precambrian plate tectonics. Elsevier, Amsterdam, 1-20.
Winter, H. de la r. (1976) A lithostratigraphic classification of the Ventersdorp succession. Trans. Geol. Soc. S. Afr., 79, 31-48.
Wright, L. A. and Troxel, B. W. (1973) Shallow-fault interpretation of Basin and Range structure, southwestern Great Basin. In: De Jong, K. A. and Scholten, R. (eds.), Gravity and tectonics. Wiley & Sons, New York, 397-407.
Zindler, A. (1982) Nd and Sr isotopic studies of komatiites and related rocks. In: Arndt, N. T. and Nisbet, E. G. (eds.) Komatiites. Allen & Unwin, London, 399-420.

THE THERMAL STRUCTURE AND THERMAL EVOLUTION OF THE CONTINENTAL LITHOSPHERE

Paul Morgan*

Lunar and Planetary Institute, 3303 NASA Road One, Houston, Texas 77058, U.S.A.

ABSTRACT

For thermal considerations, the lithosphere is defined as the outer layer of the Earth in which heat transfer is dominated by conduction. Low density continental crust caps the continental lithosphere preventing its subduction, and thus it has, in general, a long and complex thermal history. The thermal regime in the near surface zone of the continental lithosphere is commonly complex, and is responsible for much of the scatter in shallow heat flow data, the main data set used to examine the thermal structure and evolution of the continental lithosphere. A review of continental heat flow data shows that while heat flow is commonly high in areas of recent tectonic activity, there is no simple relationship between heat flow and age of tectogenesis. Low heat flow values are found in some areas of recent tectonic activity, and high heat flow values are found in some areas of older tectonism. An analysis of the parameters controlling the thermal structure of stable continental lithosphere shows that for the same heat flux into the base of the lithosphere, the main factors controlling temperatures within the lithosphere and surface heat flow are the quantity and distribution of heat producing elements within the lithosphere. In regions of recent tectonic and/or magmatic activity, perturbations in the thermal structure of the lithosphere are a function of the style and intensity of the tectonothermal disturbance. In hot spot and extensional settings, the lithosphere is thinned and its thermal gradient increased. The thermal effects at strike-slip boundaries are generally only of local importance, but can be significant when lithospheres with different thermal structures are juxtaposed. A variety of thermal phenomena are observed and predicted for zones of convergence, including low heat flow associated with subduction, high heat flow associated with magmatic activity, and thermal inversions associated with underthrusting. In seventeen continental areas, sufficient data are available to estimate the contribution of upper crustal heat production to the surface heat flow. This contribution is highly variable, and does not appear to

*Present Address: Dept. of Geosciences, Purdue University, West Lafayette, IN 47907, USA

correlate with crustal age for Proterozoic and Phanerozoic sites, but appears to be generally lower and less variable in Archean sites. In contrast, the component of heat flow from below the upper heat producing layer, the reduced heat flow, is relatively uniform, around 27 mW m^{-2} for sites which last experienced tectonism or magmatic activity in pre-Mesozoic times, but is highly variable for younger sites. From the uniform reduced heat flow for pre-Mesozoic sites, the stable thickness of the continental lithosphere is estimated to range from 90 to 220 km, thinner lithosphere being associated with higher heat production in the lithosphere. From this thickness, thermal perturbations in the lithosphere associated with tectonothermal activity are predicted to last no more than a few 100 m.y., a prediction consistent with available reduced heat flow data. Some prolongation of thermal relaxation is expected due to post-tectonic sedimentation and erosion, but these effects appear to be negligible at sites of Early Paleozoic age and older. The scatter in surface heat flow values at these sites is primarily the result of crustal heat production variations and near surface effects. Relatively low and uniform heat flow at Archean sites perhaps reflects relatively low and uniform crustal heat production at these sites.

1. INTRODUCTION — THE CONTINENTAL LITHOSPHERE

The concept of a lithosphere with strength overlying an asthenosphere without strength was first recognized by Barrell (1914a,b) and used to explain local crustal support of surface loads, while isostatic equilibrium prevails on a larger scale. With the recognition of the importance of large scale lateral movements of the Earth's crust in the framework of plate tectonics, the lithosphere has been redefined as a mobile near-surface layer of strength (Isacks et al., 1968). As strength in the lower lithosphere is controlled by creep deformation, which is a complex function of a number of variables, including temperature and strain rate (e.g., Weertman, 1970), the lithosphere has different apparent thicknesses applicable to different strain rates, and is laterally variable in thickness as a function of temperature gradient. In general, the lithosphere appears thicker as the strain rate increases, and thinner as the temperature gradient increases.

A convenient definition of the lithosphere for thermal considerations is the outer layer of the Earth in which heat transfer is dominated by conduction. In this definition, it is the strength of the lithosphere that resists thermally induced buoyancy forces which drive convective heat transfer in the underlying asthenosphere. Deformation can and does occur in the thermal lithosphere (hereafter referred to simply as the lithosphere), in response to tectonic forces, which causes non-conductive heat transfer within the lithosphere. In addition, the lithosphere can be penetrated by magmas, and its thickness varies both spatially and temporally in response to changing thermal conditions. For the purposes of this paper, the continental lithosphere is defined as those portions of this lithosphere capped by continental crust which can be deformed, but are prevented from subduction by their buoyancy.

In this paper the thermal structure and evolution of the continental lithosphere are examined. As the primary constraints on this discussion are surface heat flow data, a summary of these data and the factors which modify them are presented. Heat flow from the continents has recently been discussed on a global basis by Vitorello and Pollack (1980), Sclater *et al*. (1980) and Pollack (1982). The present paper is in many respects an extension of these studies, emphasizing the diversity of thermal phenomena in the continental lithosphere, primarily in a framework of plate interactions. In most of the thermal discussion, the base of the lithosphere is treated as a discrete boundary as a matter of convenience. It is recognized, however, that this boundary is certainly gradational, and is a rheological transition. The criteria for defining the base of the thermal lithosphere, and its relationship to other definitions of the lithosphere are discussed briefly towards the end of the paper.

2. HEAT TRANSFER IN THE LITHOSPHERE

The thermal regime of the continental lithosphere is controlled by the balance between the supply of heat to its base, the heat produced or absorbed within the lithosphere, and the heat lost through its upper surface. This regime is modified by spatial variations in thermal properties within the lithosphere, temporal variations in its heat gain and loss, and deformation of the lithosphere. Heat is transported within the lithosphere by conduction and convection, generated by radioactive heat production, exothermic chemical reactions and phase changes, and frictional heating, absorbed by endothermic chemical reactions and phase changes, and stored by changes in temperature. These thermal processes can be related by thermal energy conservation equations (e.g., Jaeger, 1965; Lachenbruch and Sass, 1977). It is impossible to write general solutions to these thermal equations to define the thermal regime of the lithosphere, due to the multiplicity of processes in and properties of the continental lithosphere. Solutions can be written for specific processes and limited property variations, however, and as temperature solutions can be superposed, it is generally possible to add the effects of different thermal processes to simulate complex geological processes.

It is convenient to divide the lithosphere into three zones, which may overlap in time and space, for discussion of thermal processes in the lithosphere. In the upper, near-surface zone, temperatures can be strongly affected by near-surface processes, which must be taken into account in the measurement and evaluation of surface heat flow. The thermal structure of the middle, internal zone of the lithosphere responds to the heat balance and thermal properties of the lithosphere, which define its steady state thermal structure. Internal deformation and magmatic intrusion within this zone, and interaction between the lithosphere and asthenosphere in the lower boundary zone of the lithosphere cause transient thermal disturbances in the lithosphere.

3. THE SHALLOW THERMAL REGIME AND THE MEASUREMENT OF HEAT FLOW

In the near-surface zone temperatures are affected by surface temperature variations, surface geological processes such as erosion and deposition and near-surface geologic structure, groundwater circulation, and near-surface heat production and absorption. The effects of these processes on subsurface temperatures have been discussed by Jaeger (1965), Roy *et al*. (1972), Kappelmeyer and Haenel (1974), Lachenbruch and Sass (1977), Blackwell (1978), Pollack (1982), and others. Although this portion of the lithosphere is not the most important in terms of the gross thermal properties of the lithosphere, it is the zone in which our only direct thermal measurements are made, and must be well understood if shallow heat flow data are to be used to compute temperatures at depth. Estimates of the effects of the shallow thermal regime are usually applied as corrections to measured heat flow data. Heat flow measurements represent only the shallow conducted component of heat flow, and these data can be supplemented by estimates of non-conductive heat loss, such as discharge of thermal springs or volcanic eruptions (e.g., Blackwell, *op cit*), and other indirect petrologic or geophysical indicators of subsurface temperature (e.g., Seager and Morgan, 1979).

A. Measurement of Heat Flow

In general it is practical only to measure the conducted component of heat flow $\hat{q}$ defined by

$$\hat{q} = -K \nabla T, \qquad (1)$$

where K is thermal conductivity and T temperature. Thus, a measurement of both thermal gradient, ∇T, and conductivity, K, are required for a heat flow determination. On land thermal gradient measurements are usually made in vertical boreholes, with laboratory thermal conductivity determinations on samples of the formations penetrated by the boreholes. Only the vertical component of heat flow is determined. If gradient measurements are made in an inclined borehole, or from temperature measurements at different depths in a mine, it is common to correct these data to yield the vertical temperature gradient. By convention, although tempertures generally increase with depth, the sign in the conduction equation is reversed, and the vertical component of heat flow, or simply heat flow, q, is given by

$$q = K\frac{\partial T}{\partial z}, \quad (2)$$

where $\partial T/\partial z$ is the vertical temperature gradient with z positive downwards. Generally, K is not constant with depth, and if the heat flow is constant the gradient changes in an inverse relationship to the conductivity. If K varies significantly with depth, a single heat flow value can be calculated by taking weighted mean gradients and conductivities, or by measuring the slope of the temperature as a function of thermal resistance from the surface (e.g., Jaeger, 1965). Techniques for measuring thermal gradients and conductivities on land are described by Birch (1950), Beck (1965) and Sass et al. (1971a). Because of near-surface thermal effects described below, heat flow measurements are usually made in boreholes ranging from around 100 m to a few km in depth.

The units of terrestrial heat flow are $mW\ m^{-2}$ or $\mu cal\ cm^{-2}\ s^{-1}$ (sometimes called Heat Flow Units or HFU: $41.8\ mW\ m^{-2} = 1.0\ \mu cal\ cm^{-2}\ s^{-1}$), and regional heat flow values typically range from 30 to 125 $mW\ m^{-2}$. Conductivities are measured in $W\ m^{-1}\ K^{-1}$, or $mcal\ cm^{-1}\ {}^{o}C^{-1}\ s^{-1}$ ($0.239\ W\ m^{-1}\ K^{-1} = 1.0\ mcal\ cm^{-1}\ {}^{o}C^{-1}\ s^{-1}$), and typically range from 0.7 to 4.0 $W\ m^{-1}\ K^{-1}$. The geothermal gradient is measured in ${}^{o}C\ km^{-1}$, and regional gradients typically range from 10 to 50 ${}^{o}C\ km^{-1}$.

Ocean bottom temperatures are generally much more stable than land surface temperatures and heat flow determinations can be made at sea from thermal gradient and conductivity measurements in the uppermost several m of sediment (e.g., Langseth, 1965). Similar thermal conditions can exist in lake sediments, and oceanic techniques have been adapted for continental heat flow determinations in several lakes (e.g., Von Herzen and Vacquier, 1967; Sclater *et al*., 1970; Morgan *et al*., 1977). These data provide a valuable supplement to the heat flow data set where borehole heat flow determinations are unavailable or inadequate.

Indirect estimators of surface heat flow using geochemical methods have also been developed, in particular a technique based on the content of silica in groundwater (Swanberg and Morgan, 1979, 1980). This technique uses an empirical relationship between the temperature dependence of quartz solubility in groundwater and heat flow, and relies upon statistically large data sets in which the depths to which the groundwaters circulate and the shallow thermal conductivities average mean "global" values. In certain groundwater and conductivity regimes, the predictions of the silica/heat flow technique can be significantly in error, and it should not be used in areas where conventional data are not available for consistency checks. It is a valuable technique for interpolating between widely spaced heat flow measurements, and can often be applied using existing groundwater data.

B. Near Surface Thermal Regime

In the near surface zone, heat flow from depth is modified by a variety of steady-state and transient thermal phenomena. Steady state thermal perturbations are introduced by lateral components of heat flow due to variations in ground surface temperature (e.g., Birch, 1950; Roy *et al*., 1972; Blackwell *et al*., 1980), and lateral subsurface variations in thermal conductivity, and conductivity anisotropy (e.g., Simmons, 1961; McBirney, 1963; Jaeger, 1965; Lee and Henyey, 1974). Quasi-steady-state perturbations may be caused by steady groundwater flow, either associated with young intrusions (e.g., Sorey and Lewis, 1976; Morgan *et al*., 1977), or larger scale flows driven by regional piezometric gradients (e.g., Brott *et al*., 1981; Morgan *et al*., 1981). Quasi-steady-state perturbations may also be caused by shallow heat production, such as the exothermic oxidation of sulphides to sulphates (e.g., Lovering, 1948; Kappelmeyer and Haenel, 1974, pp. 166-168; Morgan et al., 1983).

Transient thermal perturbations are caused by changes in surface temperature or relative surface position, due to climatic effects, sedimentation or erosion, or changes in groundwater flow or heat production. Heat flow determinations are commonly made at depths greater than the 15 m or so penetration of the annual temperature cycle. Corrections may be required for longer term temperature variations, such as those associated with Pleistocene glaciations, which penetrate to greater depths, however (e.g., Birch, 1948; Allis, 1978; but see also Sass et al., 1971b). Erosion effectively moves hot rock towards the surface, increasing the thermal gradient, and sedimentation has the opposite effect. These effects may be complicated by changes in surface elevation and surface temperature, and by the removal or addition of heat producing elements (e.g., Benfield, 1949; Birch, 1950; England and Richardson, 1980). The accuracy with which the effects of erosion, sedimentation and other near-surface thermal perturbations can be estimated depends on the precision with which the parameters defining these processes, such as erosion rate, flow rate or surface temperature distribution, are known. Commonly corrections can only be estimated.

Although the thermal structure of the near-surface zone can be complex, the thermal processes in this zone, with the possible exception of deep erosion, rarely perturb the temperatures more than a few tens of degrees at a depth of a few kilometers. As these perturbations are attenuated with depth, they have little effect on the deep thermal regime of the continental lithosphere. It is a reasonable approximation that the near-surface zone can be ignored for most thermal considerations of the bulk of the lithosphere. As they can cause significant anomalies and large uncertainties in heat flow determinations under unfavorable circumstances, however, they are of major importance to our understanding of the thermal structure of the continental lithosphere.

4. THE CONTINENTAL HEAT FLOW DATA SET

The number of continental heat flow determinations has increased by more than an order of magnitude over the last two decades, from around a hundred determinations in the early sixties (Lee and Uyeda, 1965), to on the order of two thousand determinations in 1982 (D.S. Chapman and H.N. Pollack, personal communication, 1982). A few areas, such as the western United States and parts of Europe, have sufficient data for contouring, but most of the continents are still grossly undersampled. The data available from a global heat flow data file (Chapman and Pollack, op cit) updated to mid-1982 are presented and discussed briefly continent by continent below.

A. Africa

Large areas of Africa remain unsampled by heat flow measurements, but the data which exist (Figure 1) define an interesting pattern of heat flow as a function of age of the last tectonothermal event. The Precambrian Kalahari craton of South Africa is characterized by relatively low heat flow, and even lower values have been reported from the West African craton (Chapman and Pollack, 1974). In the areas that skirt the Precambrian cratons, the heat flow range is much greater, being highest in areas of late Cenozoic volcanism. In areas of late Cenozoic tectonic activity without voluminous volcanism (much of the western arm of the East African rift system), the heat flow is not significantly higher than the craton values, although these data may be locally disturbed (Morgan, 1981). In contrast, above normal values have been reported from Zambia, an area not affected by late Cenozoic rifting (Chapman and Pollack, 1975, 1977). A trend of increasing heat flow on the Egyptian Red Sea margin has been measured, which continues offshore with very high heat flow in the axial portion of the Red Sea (Morgan et al., 1980). No heat flow data are available from the areas of the mid-Paleozoic and more recent orogenies which have affected the northwestern and southeastern margins of the African continent. High heat flow appears to be dominantly associated with late Cenozoic volcanism of the East African rift system (Morgan, op cit).

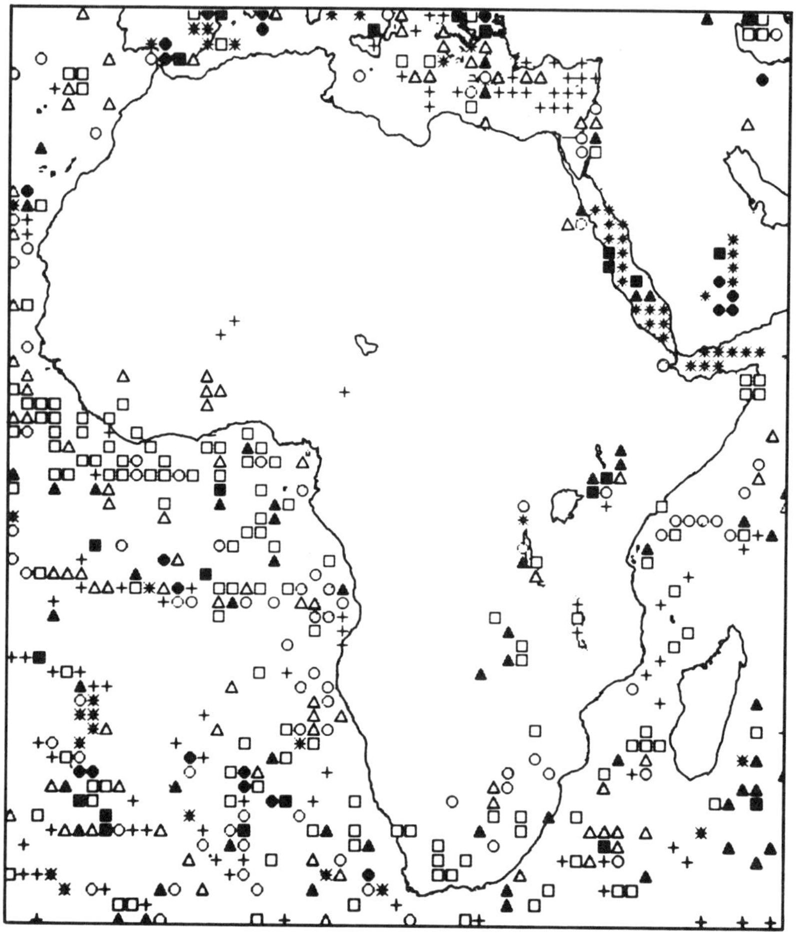

Figure 1. Heat flow data for Africa and surrounding regions. Data are averaged in 2° x 2° blocks. Key for heat flow values in $mW\ m^{-2}$: cross $\leq$ 32; open triangle 33-45; open circle 46-53; open square 54-66; filled triangle 67-78; filled circle 79-91; filled square 92-104; asterisk $\geq$ 105.

B. Antarctica

Heat flow determinations in Antarctica are very few. The only published data come from the Ross Island-Dry Valley area, and indicate moderately high to very high heat flow (Figure 2; Bucher and Decker, 1980). These high values again come from an area of late Cenozoic volcanic and extensional tectonic activity.

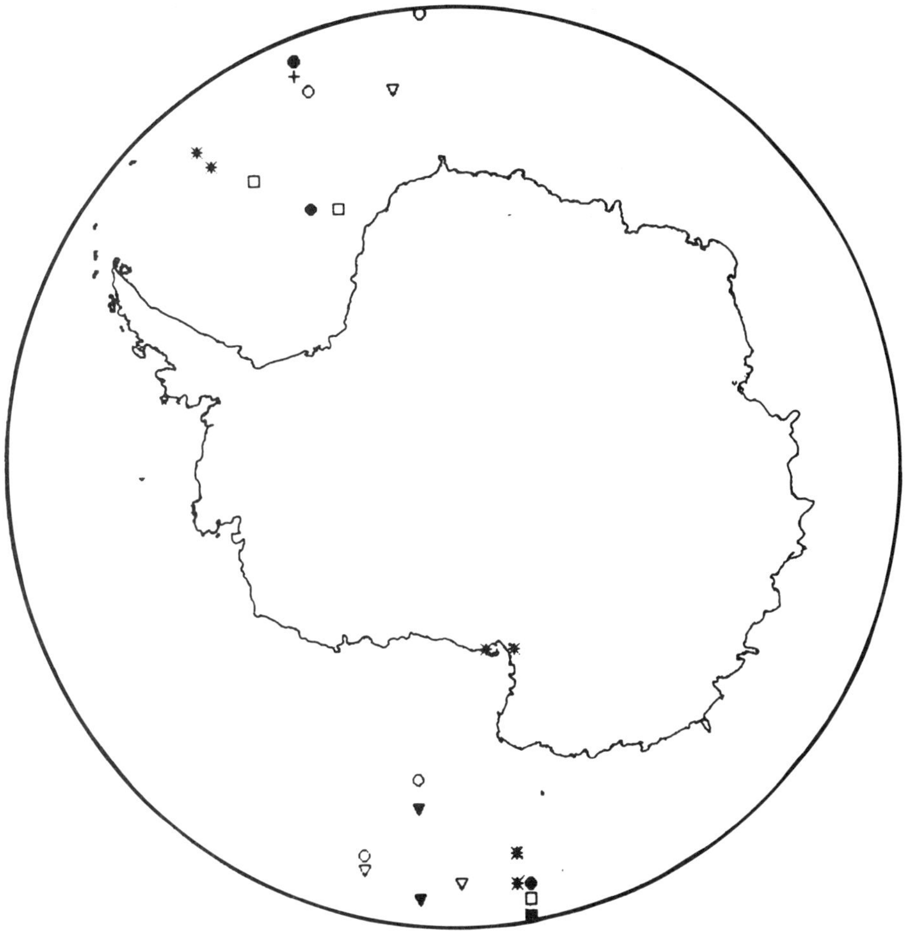

Figure 2. Heat flow data for Antarctica and surrounding regions. Data averaged and key as for Figure 1.

C. Asia

Heat flow data are sparse throughout most of the Asian continent (Figure 3). Heat flow is generally low to normal throughout the stable portions of the Soviet Union, with some higher values to the south associated with Alpine-Himalaya collisional tectonics, and high values in the Cenozoic Baikalian rift (Lubimova *et al.*, 1972). Most of the heat flow data from the Indian subcontinent are low to normal, consistent with the stable continental setting of all but the northern portion of this area (Gupta, 1982). Heat flow data from the Japanese Islands are sufficiently closely spaced for contouring (Sclater *et al.*, 1980), and indicate a reasonably consistent pattern of low heat flow on the southeastern margin of the islands increasing to high heat flow to the northwest, and into the Japanese Sea. This heat flow pattern is related to the subduction of the Pacific plate beneath the Japanese Islands, and is locally perturbed by volcanic activity.

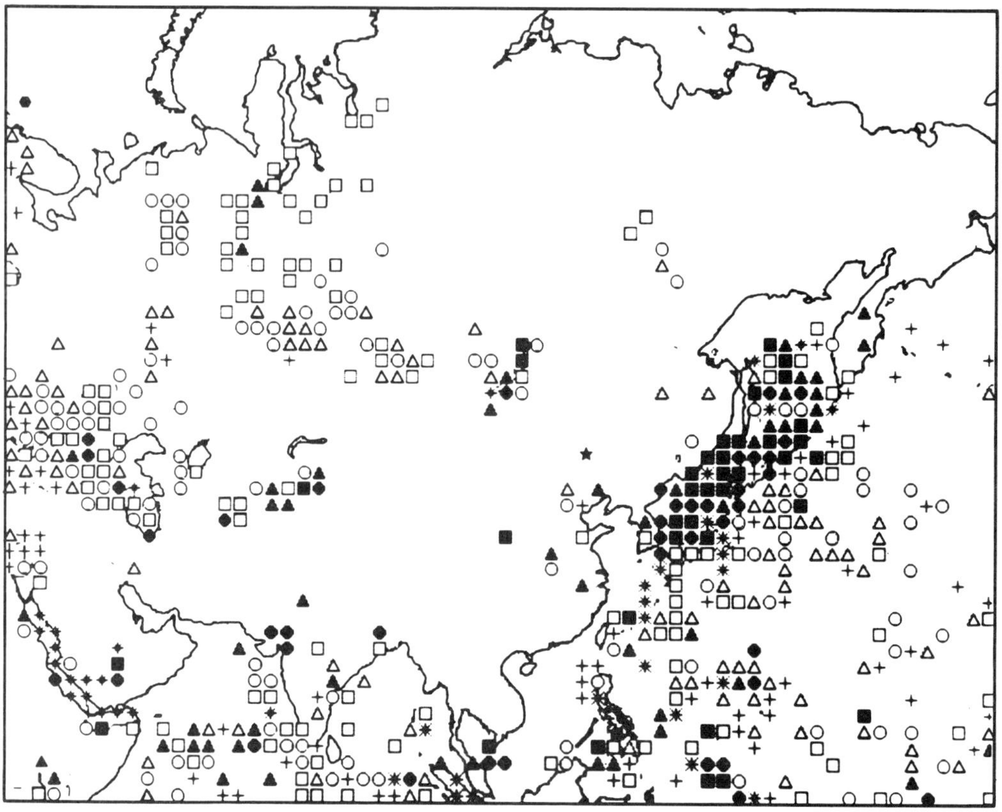

Figure 3. Heat flow data for Asia and surrounding regions. Data averaged and key as for Figure 1.

D. Australia

Approximately ninety heat flow determinations are available for the Australian continent (Figure 4), which suggest normal to low heat flow over most of the continent, characteristic of the stable tectonic setting, with the notable exception of a band of high heat flow, about 500 km wide, stretching from the Northern Territory south-southeastward to Tasmania (Sass and Lachenbruch, 1979). The highest heat flow values are in the southern portion of this band, the site of widespread Cenozoic magmatic and tectonic activity. High heat flow in the northern portion of the band does not appear to have a

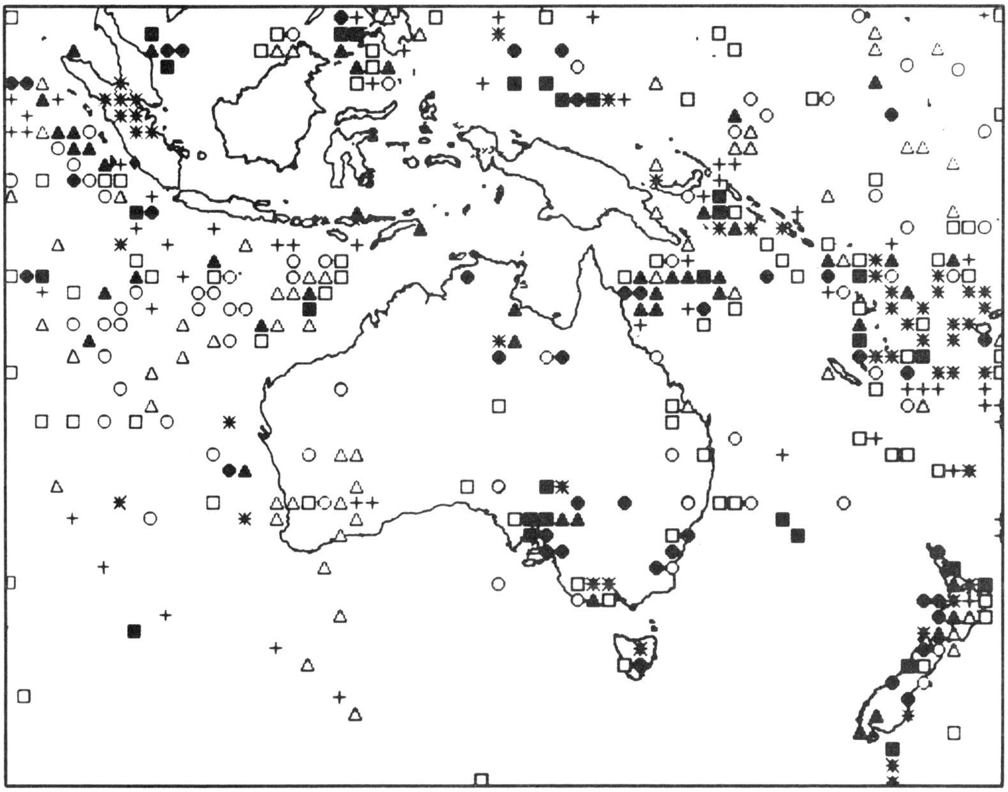

Figure 4. Heat flow data for Australia and surrounding regions. Data averaged and key as for Figure 1.

tectonic or magmatic origin. To the northeast of Australia, high heat flow values have been measured on Sumatra, in a similar tectonic setting to high heat flow on the northwestern margin of the Japanese Islands, and undoubtably related to subduction in the Sunda arc. To the southeast, very variable heat flow values have been reported from New Zealand, where heat is clearly redistributed by the hydrothermal activity, especially in North Island, but the ultimate source of high heat flow appears to be related late Cenozoic magmatic and tectonic activity.

E. Europe

The density of heat flow data in Europe is reasonable (Figure 5), and with additional subsurface temperature data, has been used to construct a heat flow map of Europe (Cermak and Hurtig, 1979), and has been the subject of a recent monograph (Cermak and Rybach, 1979). In the north, heat

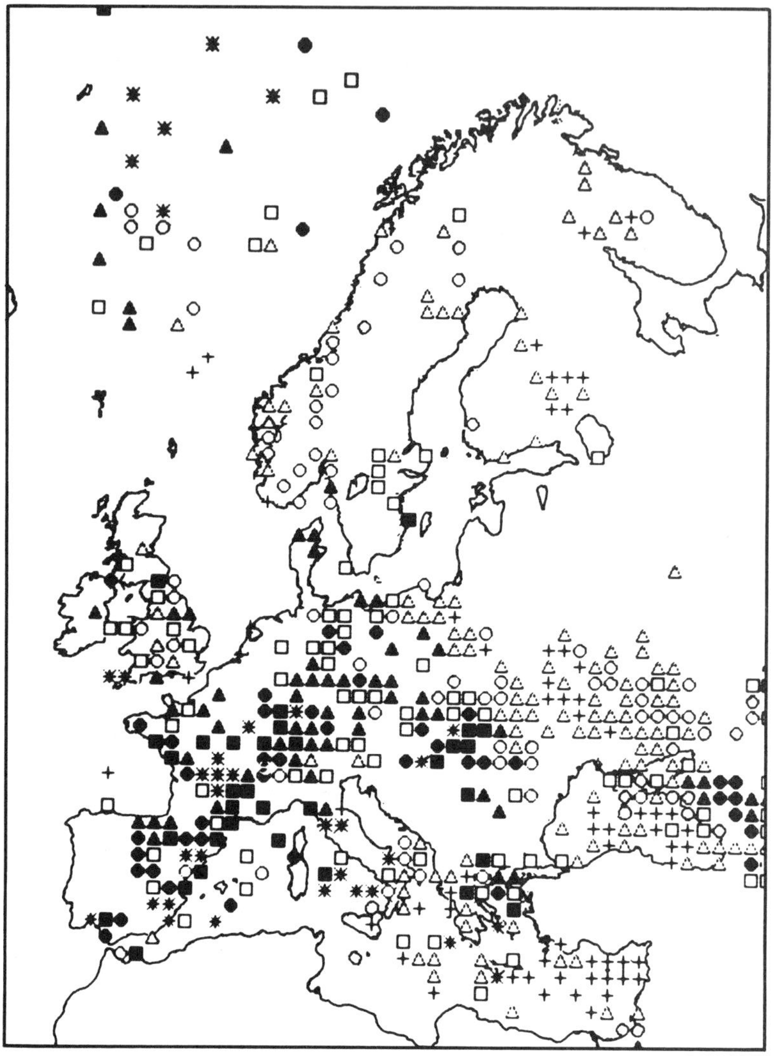

Figure 5. Heat flow data for Europe and surrounding regions. Data averaged and key as for Figure 1.

flow values from the Baltic shield are low. Further south, heat flow is much more variable, and generally higher. High heat flow values are commonly associated with Alpine tectonics, but are by no means restricted to this zone; for example, high heat flow values are reported for the Hercynian granites in southwest England (Tammemagi and Wheildon, 1974, 1977). Heat flow is high in the areas of Cenozoic tectonism and magmatism of the Rhinegraben, but is also moderately high in the adjacent areas north and west of the Alps. There is no obvious correlation between surface heat flow and crustal thickness (Cermak, 1979). A very complex heat flow pattern is shown on the heat flow map of Europe (Cermak and Hurtig, op cit), and with the exception of the low heat flow values of the Baltic shield, and very high heat flow values associated with recent magmatic activity, it is difficult to generalize about the relationship between heat flow and tectonic age for most of Europe. Undoubtably, some of the lateral variations in heat flow are due to near surface processes, and do not have a deep origin.

F. North America

A reasonable heat flow data coverage in the North American continent, especially in the western United States (Figure 6), has allowed parts of this continent to be

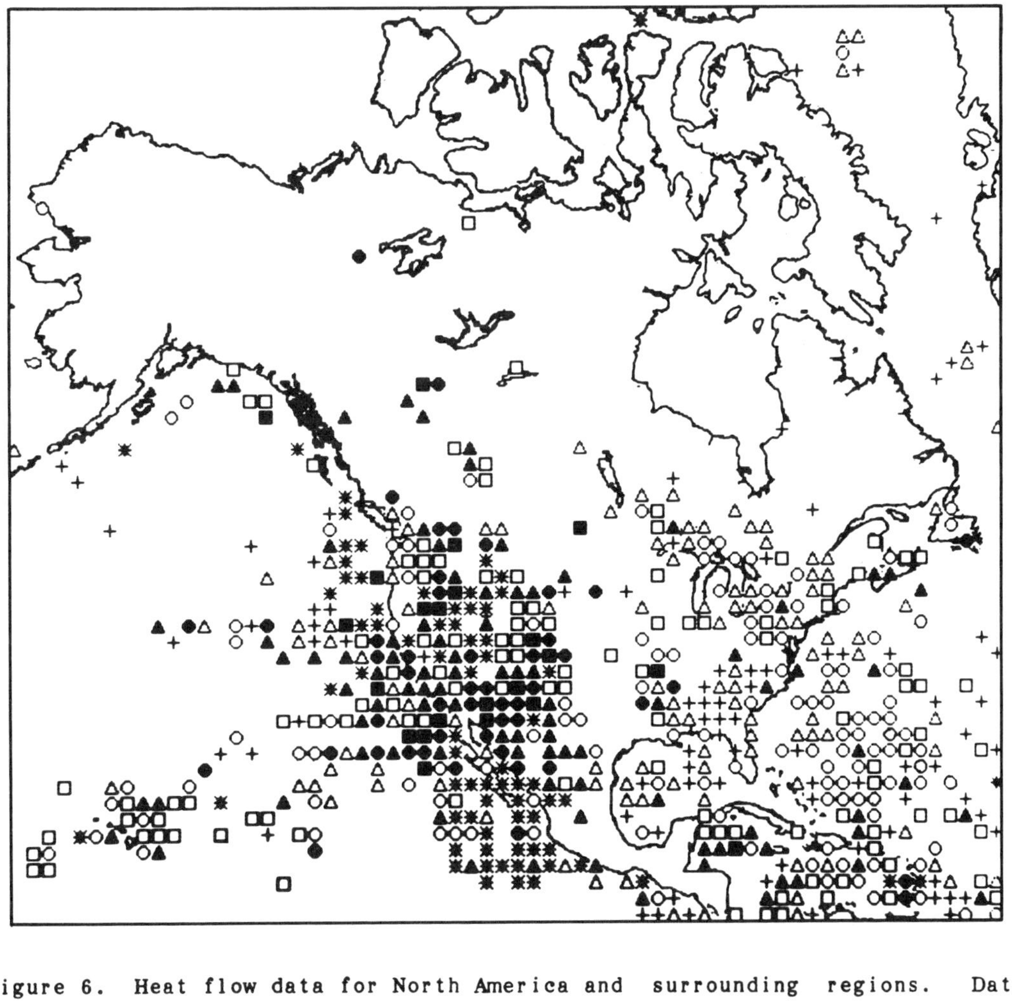

Figure 6. Heat flow data for North America and surrounding regions. Data averaged and key as for Figure 1.

contoured into thermal provinces (Roy *et al.*, 1972; Lachenbruch and Sass, 1977; Blackwell, 1978; Sass *et al.*, 1981). The eastern, stable portion of this continent is characterized by normal to low heat flow, although a significant number of data from this area indicate local thermal anomalies. In the western areas of Cenozoic tectonic activity, extensional provinces such as the Rio Grande rift and the Basin and Range are characterized by high heat flow, and these high heat flow zones appear to continue southward into the areas of Cenozoic tectonic activity in Mexico, and northward into the Northern Rocky Mountains and Columbia Plateau. Moderate to low heat flow values have been reported from the interior of the Colorado Plateau, the Sierra Nevada, and coastal regions of the Pacific Northwest. Regions of high heat flow are characterized by a large scatter in the data, indicating redistribution of heat by hydrothermal circulation.

G. South America

Much of the South American continent is unsampled by heat flow data (Figure 7), but some data are available from the eastern and western portions of the continent. The cratons and intercratonic basins of eastern South America are characterized by low to normal heat flow. A more complex heat flow pattern is apparent along the western margin, where low heat flow values have been reported for sites close to the coast, with high heat flow at some sites further into the Andes. Interestingly, however, not all interior areas of the Andes have high heat flow. Portions of the Andes have a similar heat flow profile to the Japan arc, but data from other areas show no increase in heat flow away from the coast.

H. The Global Heat Flow Data Set

On a global basis, several authors have noted that the mean heat flux from young terrains is higher than that from older terrains (e.g., Kraskovski, 1961; Lee and Uyeda, 1965; Polyak and Smirnov, 1968; Hamza and Verma, 1969;

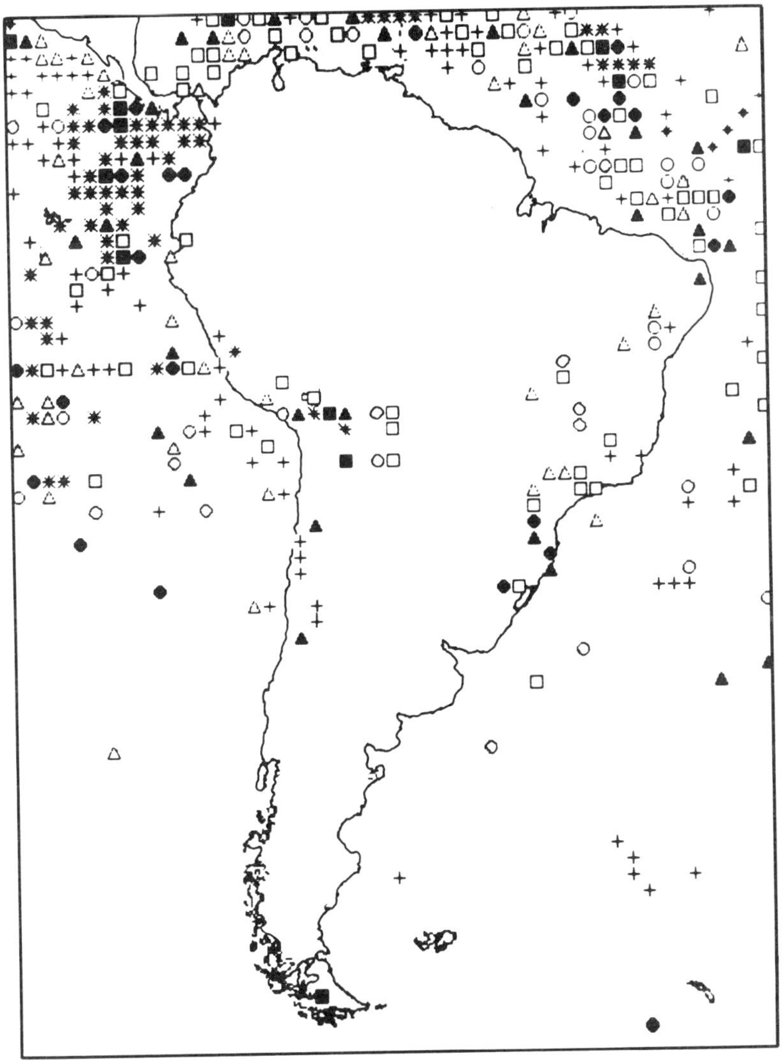

Figure 7. Heat flow data for South America and surrounding regions. Data averaged and key as for Figure 1.

Sclater and Francheteau, 1970; Chapman and Furlong, 1977; Cermak and Rybach, 1979; Sclater et al., 1980). Two recent global compilations of continental heat flow as a function of age are shown in Figure 8. These two compilations appear to show similar trends. However, there are significant differences in the details of the trends, and in the methods used to assign age to the heat flow sites.

In the first compilation (Figure 8A), Chapman and Furlong (1977, in Vitorello and Pollack, 1980) assigned ages on the basis of the last tectonothermal event, and made seven divisions of geological time. This compilation shows approximately constant mean heat flow (around 72 mW m^{-2})

at Cenozoic and Mesozoic sites, with a larger scatter in the data from Cenozoic sites. Mean heat flow for the Late Paleozoic (61 mW m^{-2}) is intermediate between the means for the younger provinces and the means for sites ranging in age from Early Paleozoic to Early Proterozoic (51-54 mW m^{-2}).

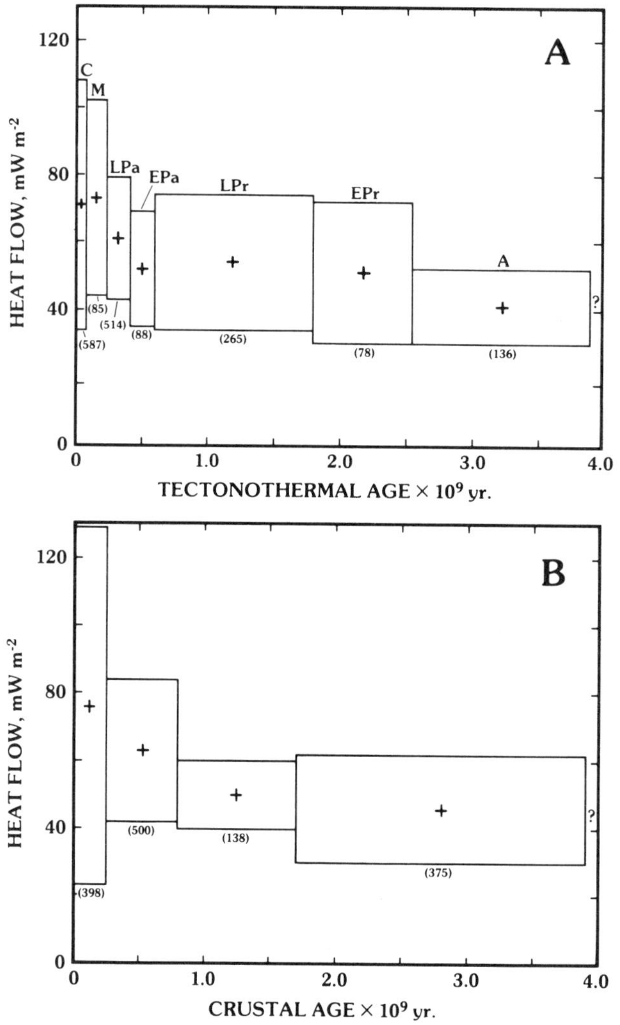

Figure 8. The global continental heat flow data set: A) heat flow data averaged in groups according to age of the last tectonothermal event at each site (after Chapman and Furlong, 1977; in Vitorello and Pollack, 1980) - age key, C Cenozoic, M Mesozoic, LPa Late Paleozoic, EPa Early Paleozoic, LPr Late Proterozoic, EPr Early Proterozoic, and A Archean; and B) heat flow data averaged in groups according to radiometric crustal age at each site (after Sclater et al., 1980). In each plot, mean heat flow values in each group are plotted as crosses at the mid-point of the age range; Boxes around the crosses indicate the age ranges for the data and standard deviations of the means. The number of data in each group are indicated by each box in parentheses.

Heat flow data from Archean sites appear to be generally lower than data from younger Precambrian sites (mean = 41 mW m^{-2}), and the data show less scatter than all data from younger sites. Although it can be argued that this compilation shows a general decrease in heat flow with age of the last tectonothermal event, it could equally well be argued that heat flow is constant for sites ranging in age from Early Paleozoic to Early Proterozoic, is generally higher for younger sites, but lower for Archean sites.

In the second compilation (Figure 8B; Sclater et al., 1980), ages were assigned on the basis of the basement age at each heat flow site, and only four age divisions were used. This method of age assignment is significantly different from that used by Chapman and Furlong (1977), as it uses the radiometric age when the crustal rocks now at the surface were last strongly heated, commonly the age of last orogenesis, and ignores later tectonothermal events which do not cause resetting of the surface rock ages. In a thermal analysis of the lithosphere, tectonothermal age assignment is preferred over crustal age, as high heat flow values from young volcanic provinces, such as East Africa and eastern Australia, are clearly related to these late magmatic events rather than the much older crustal ages. The greater age ranges used in the compilation by Sclater et al. (op cit) also appear to mask some possibly significant detail in the heat flow/age trend. The mean heat flows for each age group used in this compilation show a relatively smooth decrease in heat flow with increasing age. However, the standard deviations of these means do not show a similar decrease with age (the smallest data set having the smallest standard deviation), which indicates that different age groups may be sampling different populations of heat flow data. Again, this compilation could be interpreted as indicating a smooth decrease in heat flow with tectonic age. Alternatively, it could be argued that the data indicate a rapid decrease in heat flow for Phanerozoic sites, with approximately constant heat flow for Precambrian sites.

On a continental scale, as in the data summarized above (Figures 1-7), the generalizations apparent in the global compilations have many notable exceptions. Stable shield areas are generally characterized by low to moderate heat flow, and within any one stable province, the scatter in heat flow is generally small (a notable exception is the Central Shield Province of Australia, with high mean heat flow and large data scatter; Sass and Lachenbruch, 1979). However, different stable areas have a range in heat flow means, and this range produces some of the apparent scatter in the older age groupings of the heat flow data in the global compilations. The mean in each group represents a multi-modal data set, and thus may require more complex analysis and representation than a simple mean and standard deviation. In contrast, in individual younger tectonothermal areas, the heat flow data are commonly characterized by a large scatter. This scatter is obviously in part related to the importance of shallow thermal processes in these active tectonic areas, notably the effects of hydrothermal circulation. Heat flow is generally higher in younger terrains, although low to very low heat flow is reported for some active tectonic areas associated with subduction (e.g., Sierra Nevada; Blackwell, 1971; Roy et al., 1972). Thus at least some of the scatter about the mean heat flow in young terrains is the result again of a multi-modal data set, and is not purely random scatter about a mean. Global heat flow data compilations do not discriminate among different tectonic and thermal processes, or between different components of the surface heat flow. To investigate the importance of these processes and components, it is necessary to examine the thermal structure and tectonic processes within the lithosphere in more detail.

5. STEADY-STATE THERMAL STRUCTURE OF THE LITHOSPHERE

Lithospheric heat transfer is primarily conductive except during periods of active tectonism and magmatism when convection may dominate. In this section steady-state heat flow in the lithosphere is examined, assuming the upper and lower boundaries of the lithosphere to be stable boundary conditions, and no deformation or magmatic activity in the lithosphere.

Lateral variations in thermal properties within the lithosphere are greatest at shallow depths, and are of minor importance to the deep thermal structure of the lithosphere. If only vertical property variations are considered, the heat flow problem reduces to one dimension, temperature T at depth z is given by the solution of

$$\frac{d}{dz}\left(K\frac{dt}{dz}\right) = -A \tag{3}$$

(Jaeger, 1965) where K and A are thermal conductivity, and heat production per unit time per unit volume, respectively, both of which are, in general, a function of depth. If heat production is negligible, equation (3) reduces to equation (2), and under steady-state conditions, i.e., heat flow = constant, temperature gradient is inversely proportional to thermal conductivity, and thus temperature at depth is sensitive to changes in thermal conductivity above the point of interest.

Thermal conductivity depends on a number of factors including temperature, mineralogy and porosity, and data for selected rocks have recently been compiled by Roy *et al.* (1981). Earth materials in the lithosphere can have a wide range in thermal conductivity at surface temperatures, ranging from $<1\ W\ m^{-1}\ K^{-1}$ for unconsolidated sediments to $>6\ W\ m^{-1}\ K^{-1}$ for halite and quartzite. Most of this variation comes from the effects of low conductivity fluids filling pore space and high conductivity monomineralic rocks both typically found at relatively shallow depths. Crystalline rocks likely to constitute a significant volume of the crust and upper mantle have a smaller range in mean thermal conductivities at temperatures up to 1000 ^{o}C, 2.0 to 3.5 $W\ m^{-1}\ K^{-1}$, but this range is significant, and it is necessary to construct a thermal conductivity model with which to calculate temperatures at depth. For example, Smithson and Decker (1974) divide the continental crust into three zones based upon seismic and thermal properties, and assign a thermal conductivity of 2.7 W m^{-1} K^{-1} to the upper and middle zones, a conductivity of 2.1 W m^{-1} K^{-1} to the lower zone and a conductivity of 2.9 W m^{-1} K^{-1} to the upper mantle. Fortunately the range in probable thermal conductivities for bulk crustal and uppermost mantle rocks is less than a factor of two, but without good control the exact nature of these rock types, any lithospheric thermal conductivity model is necessarily somewhat subjective.

In the above discussion, the effects of temperature and the mode of conduction have been largely ignored as in the crust, at temperatures less than 1000 ^{o}C, composition is probably the most important factor controlling conductivity. Below around 400 ^{o}C lattice, or phonon conductivity, which decreases with increasing temperature, is believed to be the main mechanism for heat transport. Above this temperature, radiative, or photon conductivity becomes more important, and the trend of decreasing conductivity with depth is reversed (e.g., Birch and Clark, 1940; Clark, 1957; Fujisawa et al., 1968; Kanamori et al., 1968; Schatz and Simmons, 1972; MacPherson and Schloessin, 1982). At any temperature, the effective conductivity, K, is given by the sum of the lattice and radiative conductivities, K_l and K_r, respectively. On the basis of experimental results, for a hypothetical mantle of olivine composition $Fo_{90}Fa_{10}$, Schatz and Simmons (op cit) give the following relationships for K_l and K_r as a function of absolute temperature, T_K:

$$K_l = (74 + .50T_K)^{-1} \text{ W m}^{-1} \text{ K}^{-1} \tag{4}$$

$$K_r = 0 \qquad T_K \leq 500 \text{ K}$$

$$K_r = 2.3\times10^{-6}(T - 500) \text{ W m}^{-1} \text{ K}^{-1} \qquad T_K > 500 \text{ K} \tag{5}$$

where K_1 has a minimum value K_{1min} given by

$$K_{1min} = 1.26(1 + z \times 10^{-6})\ W\ m^{-1}\ K^{-1} \tag{6}$$

where z is depth in m. To illustrate the effects of these relationships, conductivity and temperature are shown as a function of depth in Figure 9, constructed assuming a uniform heat flow throughout the section of 25 mW m^{-2}, and that the above relationships apply to the surface. Also shown for comparison in Figure 9 are the temperature profiles for the same heat flow assuming constant thermal conductivities of 2.5 and 3.4 W m^{-1} K^{-1}, as used by Sclater et al.(1980). The maximum departure of the uniform 3.4 W m^{-1} K^{-1} model temperatures from the Schatz and Simmons model temperatures is 100 ^{o}C, and in view of the uncertainties in mantle composition, the assumption of a uniform conductivity in the mantle lithosphere appears to be reasonable. Thus, if thermal conductivity models are used primarily for comparisons of thermal structure in the lithosphere between one area and another, the uncertainties in thermal conductivity are of second order importance.

Uncertainties in lithospheric thermal structure are also caused by uncertainties in the distribution of heat producing elements in the lithosphere. As with conductivity, some model for the distribution of heat production with depth must be used to solve equation (3) for temperature as a function of depth. Unlike conductivity, however, which is a bulk property of a rock, most heat production is a function of uranium, thorium and potassium concentrations in the rock, specifically concentrations of the isotopes U^{235}, U^{238}, Th^{232}, and K^{40}. These concentrations cannot be reliably estimated from rock type, but as the heat producing elements are incompatible elements, they are concentrated in low melting-temperature rocks. Radiogenic heat production is generally higher in rocks of granitic composition than more basic rocks, and is significant in most continental upper crustal crystalline rocks. Many surface rocks have radiogenic heat productions on the order of

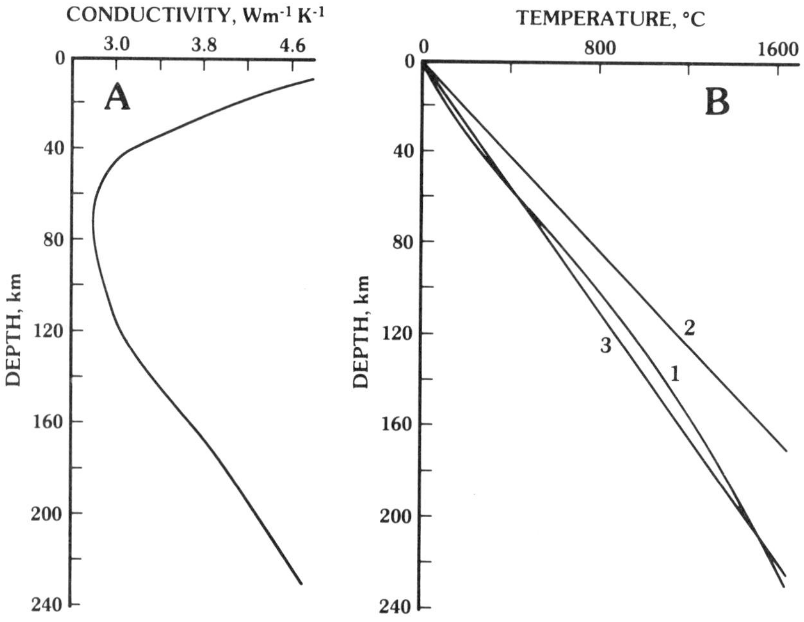

Figure 9. A) Thermal conductivity as a function of depth according to the relationships of Schatz and Simmons (1972), assuming a constant heat flow through the section of 25 mW m^{-2}. B) Temperature as a function of depth assuming a constant heat flux of 25 mW m^{-2}, and 1) conductivities as shown in A), 2) a constant thermal conductivity of 2.5 W m^{-1} K^{-1}, and 3) a constant conductivity of 3.4 W m^{-1} K^{-1}.

2 μW m^{-3} (e.g., Heier and Rogers, 1963; Blackwell, 1971), which if uniform throughout a 30 km thick crust, even with no heat flow from the mantle, would yield a surface heat flow of 60 mW m^{-2}, about the average continental heat flux. As it is unlikely that the continental mantle heat flux is zero, and there is no general correlation between heat flow and crustal thickness, heat production must decrease at depth.

Two primary components to heat flow measured in stable continental areas are generally acknowledged: Heat flow from the upper mantle and heat generated by radiogenic heat production in the crust (Birch *et al.*, 1968; Roy *et al.*, 1968; Lachenbruch, 1968). The key to separating these components from heat flow determinations in plutonic rocks was the discovery by Birch *et al.* (*op cit*) of a linear relationship between heat flow and surface heat production of the form:

$$q_s = q^* + A_s b \tag{7}$$

where q_s is the measured heat flow, A_s is the heat production of the plutonic rocks in which the measurement was made. q^* and b are the constants of the linear relationship, interpreted to be the heat flow from below the heat producing layer, commonly known as the reduced heat flow, and a depth scaling parameter for the heat producing layer, respectively. This discovery led to the definition of a heat flow province by Roy et al (1968) as an area in which a single relationship of the form of equation (7) was valid, and three provinces were identified in the United States, the eastern United States, the Basin and Range province, and the Sierra Nevada (Table 1). Roy et al. (1972) later redefined the term heat flow province to be a region with the same heat flow from the mantle (q^*) in recognition that this parameter was of greater geophysical significance than the distribution of heat producing elements in the crust. Subsequently, the linear heat flow/heat production relationship has also been established for sites in metamorphic rocks (e.g., Richardson and Oxburgh, 1978), and heat flow provinces have now been defined in seventeen continental areas, as listed in Table 1.

The depth parameter, b, defined in equation (7) may have some significance in defining the relative depth distributions of heat production among heat flow provinces, but does not uniquely define the form of these distributions. Under the assumption that heat production does not increase with depth at any level, two extreme distributions of heat producing elements have been proposed which satisfy equation (7). In the first, the step model, the heat production is assumed to remain constant from the surface to a depth given by b (Birch et al., 1968). In the second, the exponential model, heat production is assumed to decrease exponentially from the surface (Lachenbruch, 1968), such that heat production A_z at depth z is given by:

$$A_z = A_s \exp(-z/b). \quad (8)$$

This latter relationship has the advantage that the linear relationship between heat flow and heat production is preserved during differential erosion (Lachenbruch, *op cit*), and is consistent with heat generation data from batholiths with differing depths of emplacement (Swanberg, 1972). Costain and Glover (1979), however, have recently published data from the southern Appalachians of the USA which they interpret to support the step model. As discussed by Blackwell (1971), the actual distribution is likely to be much more complex than either model.

Recent attempts to understand the distribution of heat producing elements with depth have focussed on the processes of transport of uranium, thorium and potassium in the crust, and indicate that there may be different depth parameters controlling the distributions of these three elements in the crust, and/or different distributions at different depths (e.g., Costain, 1978; Gosnold and Swanberg, 1980; Jaupart *et al*., 1981). These transport processes include primary differentiation of the crust, magmatic and metamorphic fluid circulation, and meteoric water circulation. Thus, if the degree of crustal metamorphism is sufficiently high, the transport processes for heat producing elements are similar during major batholith emplacement and regional metamorphism, a hypothesis which appears to be corroborated by heat flow/heat production data from plutonic and metamorphic sites. Studies of uranium in metamorphic rocks indicate that it is essentially immobile up to amphibolite grade, but transported through migrating fluid phases during granulite metamorphism (Dostal and Capedri, 1978). Presumably thorium is mobilized at a similar metamorphic grade (McLennan and Taylor, 1980). Unfortunately, few heat production data exist for large vertical sections of the continental crust, and the different models cannot be rigorously tested.

Heat production data have recently been published by Nicolaysen *et al*. (1981) for a 16 km, originally vertical

TABLE 1. Heat flow, Heat Production and Age Parameters for Several Heat Flow Provinces

Province	Age 1)	Age 2)	q_s	q^*	q_A	b	n	Code	Ref
			North America						
Basin and Range	C	C	-	59±14	-	9.4	12	BR	1
ditto	C	C	92±33	69±34	23	(10.0)	86	BR'	2
Southeastern Appalachians	Pl	Pl	49±11	28±1	21	7.2	3	SEA	3
Sierra Nevada	M	C	37±13	18±3	19	10.1	10	SN	2
Eastern United States	Pr	Pr	57±17	33±4	24	7.5	15	EUS	4
Superior	A	A	34±8	21±1	13	14.4	11	SP	5
ditto, corrected	A	A	-	28±1	-	13.6	11	SP'	5
			South America						
Brazilian coastal shield	LPr	LPr	56±15	28±7	28	13.1	4	Bz	6
			Europe						
Baltic shield	APr	APr	36±8	22±6	14	8.5	3	Ba	7
Bohemian Massif	CM	C	73±18	44±1	29	3.8	2	BM	8
Ukraine	A	A	37±8	25±2	12	7.1	12	U	9
England and Wales	Pl	Pl	59±23	23±2	36	16.0	10	EW	10
			Africa						
Niger	LPr	LPr	20±8	11±8	9	(7.5)	-	N	11
Zambia	LPr	LPr	67±7	40±6	27	(7.5)	-	Z	12
ditto	LPr	CM	67±7	40±6	27	(7.5)	-	Z'	12
			Australia						
Western	A	A	39±8	26±8	13	4.5	9	WA	13
Central	EPr	EPr	83±21	27±6	56	11.1	10	CA	14
Eastern	Pl	C	72±27	57±22	15	(11.1)	-	EA	14
			Asia (Indian Shield)						
Late Proterozoic	LPr	LPr	71±11	38±2	33	14.8	4	I1	15
ditto	LPr	CM	71±11	38±2	33	14.8	4	I1'	15
Archean Shield	A	A	49±8	33±1	16	7.5	2	I2	15

Ages assigned to each province are 1) crustal age, and 2) age of last tectonothermal event (ages estimated from Condie, 1976, Figures 5-10, page 110, Sclater *et al*., 1980, and as discussed in text); Age key - C Cenozoic, M Mesozoic, CM Cenozoic-Mesozoic, Pl Paleozoic, Pr Proterozoic, LPr Late Pr, EPr Early Pr, A Archean, APr Archean-Proterozoic. q_s is mean surface heat flow (mW m^{-2}) ± standard deviation. q^* is reduced heat flow (mW m^{-2}) ± uncertainty. q_A is the estimated crustal heat production (mW m^{-2}), calculated by $q_A = q_s - q^*$. b is the depth parameter (km) for upper crustal heat production - see text. Assumed b values are given in parentheses. n is the number of heat flow/heat production data used to define the heat flow province. Code identifies the data points plotted in Figures 17 and 18. References: 1 - Roy *et al*. (1968) see also Blackwell (1978); 2 - Lachenbruch and Sass (1978); 3 - Costain (1977); 4 - Roy *et al*. (1968); 5 - Jessop and Lewis (1978); 6 - Vitorello (1978); 7 - Rao and Jessop (1975); 8 - Cermak (1975); 9 - Kutas (1977); 10 - Richardson and Oxburgh (1978); 11 - Chapman and Pollack (1974); 12 - Chapman and Pollack (1977); 13 - Jaeger (1970), Sass *et al*. (1976); 14 - Sass *et al*. (1976);

section of Archaean granitic crust, exposed in the core of an updomed and overturned sequence of strata in the Vredefort structure in South Africa. Individual heat production values for uranium, thorium and potassium are plotted as a function of depth in Figure 10. It is clear that the data do not follow simple, nor the same distributions with depth, but the uranium and thorium data can be approximated by exponential curves, as shown in the figure. Potassium does not vary significantly with depth. Other distributions, such as the more complex exponential distribution proposed by Gosnold and Swanberg (1980), could also be fitted to the data. The distribution of heat production in the crust is clearly complex, and additional data are required before generalizations about the distribution can be made.

Allis (1979) and Rybach and Buntebarth (1982) have proposed relationships among density, seismic velocity, heat generation and mineralogical constitution from which it may be possible to predict heat production from other physical properties. From these relationships, Allis (op cit) has proposed that an exponential decay in heat producing elements is only applicable to the upper crust, and that the lower crust is uniformly depleted with a mean heat production of 0.2 ± 0.1 μW m^{-3}. With the data available at present, the only possible conclusion is that the distribution of heat producing elements in the upper crust appears to be bracketed by the step and exponential models. If a different distribution is applicable to the lower crust, it could be undetectable in the available heat flow/heat production data, but is potentially significant to the thermal structure and evolution of the lithosphere, and will be discussed in a later section.

Heat production in the crust increases the shallow geothermal gradient, and increases the temperature at depth within the lithosphere. The magnitude of this temperature increase is dependent on the vertical distribution of heat production, and at depth ($z \gg b$) is twice as great for the exponential distribution as for the step distribution, for

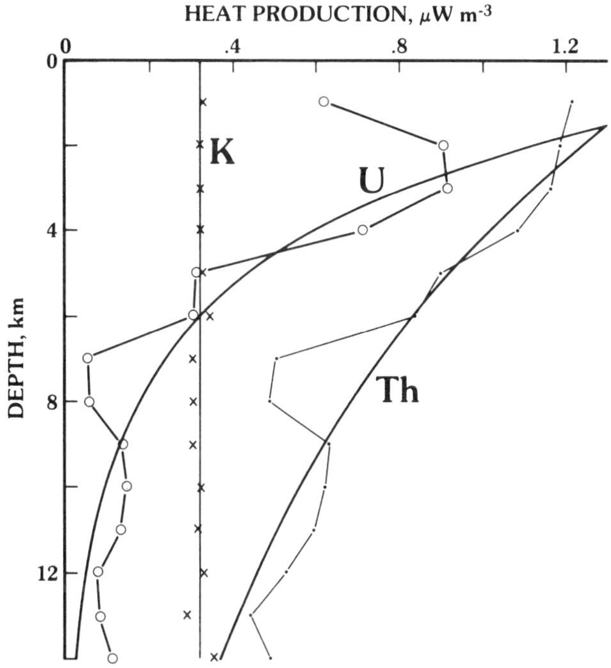

Figure 10. Heat production as a function of depth in the Vredefort structure, South Africa (data from Nicolaysen _et al._, 1981). Actual data are shown by open circles for Uranium, dots for Thorium, and x's for Potassium. Smooth curves show the results of exponential regression analyses of the Uranium and Thorium versus depth data, and the mean of the Potassium data. Statistics and parameters for these analyses are given by the curves. (Note $b_{Th} > b_U$, in contrast to $b_{Th} < b_U$ determined by Jaupert _et al._, 1981, from an analysis of heat production data from multiple sites.)

the same total heat production (Blackwell, 1971). For low surface heat production (A_s) and/or a small depth parameter (b), the distribution of heat producing elements is unimportant. With typical heat production parameters, however, A_s = 2.0 W m^{-3} and b = 10 km, and a crustal thermal conductivity of 2.5μW m^{-1} K^{-1}, the temperature increase predicted at depth ranges from 40 ^oC for the step model to 80 ^oC for the exponential model. For high heat production, A_s = 5.0 μW m^{-3} and b = 15 km, the predicted temperature increases for the two models are 225 ^oC and 450 ^oC, respectively. The thermal structure of the continental lithosphere is sensitive to both quantity and distribution of heat producing elements.

Under conditions of steady-state conduction, the thermal structure of the lithosphere depends on three factors; the thermal conductivity structure of the lithosphere, the quantity and distribution of heat producing elements within the lithosphere, and the heat flow into the base of the lithosphere. With assumptions of rock type, reasonable estimates can be made of the conductivity within the lithosphere. In heat flow provinces defined by heat flow/heat production data sets, the heat input into the lower lithosphere can be estimated by equating it with the reduced heat flow for the province. Even with conductivity and basal heat flux well defined, however, some uncertainty in temperature at depth still remains, especially in areas with high crustal heat production, as the distribution and quantity of heat producing elements in the crust cannot be uniquely defined.

6. DYNAMIC THERMAL REGIMES IN THE LITHOSPHERE

Correlations of heat flow with age since last tectonothermal or orogenic event reviewed above show a general decrease in heat flow with tectonic age. This global correlation, however, masks the subtleties in the temperature field in dynamic lithospheric processes. Although the gross thermal effects of many tectonic processes may be similar, their details differ widely and are the key to understanding the processes.

A. Extensional Tectonics and Thermal Thinning of the Lithosphere

Extensional tectonics in the continental lithosphere are manifested at the surface by rift systems, which range from relatively well defined grabens, such as the classical Gregory, or Kenya rift of the East African rift system (Baker *et al.*, 1972), to broad regions of extensional tectonics, the most extensive modern example being the North American Basin and Range province (Stewart, 1978). Universal features of rift zones are crustal (lithospheric) extension with graben formation, and anomalous crust and upper mantle, usually interpreted to be the result of thermal or mechanical lithospheric thinning or asthenospheric diapirism (Morgan and Baker, 1983). Two end-member processes for rifting have been suggested, active rifting, in which rifting is a result of thermal upwelling of the asthenosphere, and passive rifting, in which rifting is a passive response to a regional extensile stress field (Sengör and Burke, 1978; Baker and Morgan, 1981). The former process implies a thermal thinning of the lithosphere, which appears to occur both with or without rifting (as in anomalous mid plate swells and/or volcanism), and thus thermal thinning and active rifting are treated as a single thermal process.

In active rifting or thermal thinning of the lithosphere there are two modes in which thermal upwelling of the asthenosphere can affect and thin the overlying lithosphere: 1) the lithosphere can be heated by heat transfer from the asthenosphere, the base of the lithosphere being heated above the transition temperature from lithosphere to asthenosphere and effectively converted into asthenosphere, or 2) thermally induced, density driven diapirism can mechanically deform the lithosphere. In practice, probably a combination of thermal and mechanical thinning occurs, although models of diapirism (Bridwell and Potzick, 1981; Mareschal, 1983; Neugebauer, 1983) require low lithospheric effective viscosities for significant rates of diapiric ascent. This viscosity requirement indicates that the lithosphere must be preheated prior to diapirism, or that the rate of diapiric thinning is controlled by heat transfer into the lithosphere ahead of the diapir, and that diapiric ascent may be conductively controlled. Thus, for this discussion, only conductive heat transfer is considered.

Models of thermal thinning of the lithosphere by conduction of heat through its base have recently been presented by a number of authors (Figure 11; Birch, 1975; Crough and Thompson, 1976a; Crough, 1978; Gass et al., 1978; Sandwell, 1982; Spohn and Schubert, 1982, 1983; Wendlandt and Morgan, 1982; Meng and Morgan, 1982; Morgan, 1983a). The data with which these models can be directly tested are impossible to collect, but Wendlandt and Morgan (*op cit*) have interpreted petrologic data to give a record of lithospheric thinning in East Africa, and show that this thinning record is consistent with an increase in temperature gradient at the base of the lithosphere of 55 $^{o}C\ km^{-1}$. It is not suggested that the lithosphere/asthenosphere interaction during lithospheric thinning is a purely conductive process, but it does appear that the thermal effects of this interaction in the unthinned lithosphere can be modelled by assuming conductive heat transfer at the base of the lithosphere (Meng and Morgan, *op cit*).

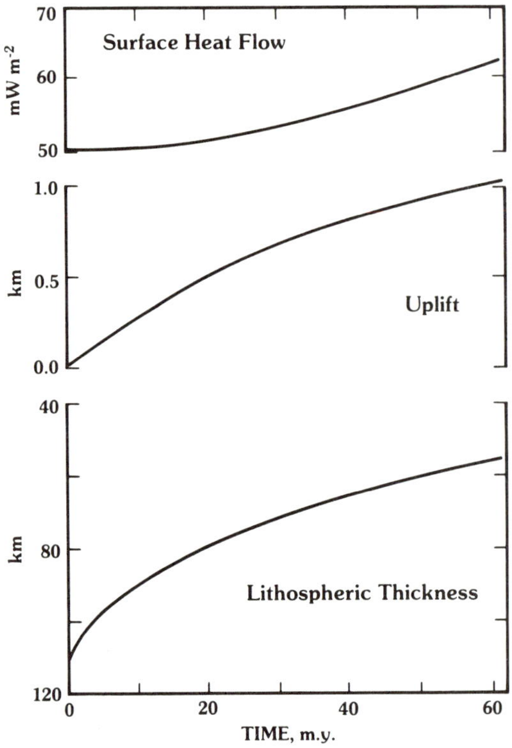

Figure 11. Thermal thinning of the lithosphere with corresponding uplift and surface heat flow. Model calculated using the moving plane source method of Morgan (1983a), assuming initial surface heat flow of 50 $mW\ m^{-2}$, surface heat production of 2 $\mu W\ m^{-3}$, heat production depth parameter of 10 km, with heat production depth distribution an average of the exponential and step models, crustal thickness of 40 km, crustal thermal conductivity of 2.5 $W\ m^{-1}\ K^{-1}$, volume coefficient of thermal expansion of $3\times10^{-5}\ K^{-1}$, lithosphere/asthenosphere boundary temperature of 1200 ^{o}C, mantle thermal conductivity of 3.4 $W\ m^{-1}\ K^{-1}$, initial lithosphere thickness of 115 km, and increase in heat flux at the base of the lithosphere of 60 $mW\ m^{-2}$. These parameters are considered to be appropriate to the Cenozoic uplift of the U.S. Colorado Plateau.

During thermal thinning of the lithosphere the response of the thermal structure of the lithosphere to the thermal perturbation at its ascending base is dependent on the rate of lithospheric thinning, the instantaneous thickness of the lithosphere, and its thermal diffusivity. Morgan (1983a) defined rates of thinning in terms of this response, with two end-members; 'rapid' thinning in which the lithosphere thins more rapidly than the thermal perturbation can propagate from its ascending base, and 'slow' thinning in which thinning is so slow that the lithosphere temperatures relax to quasi-steady-state values during the thinning. Obviously these two thinning rate criteria are only completely satisfied by an infinite and zero thinning rate, respectively. However, by considering parameters which are relevant to the measurable response of the lithosphere to thermal thinning, the mean temperature of the lithosphere which is related to thermal uplift of the lithosphere, and surface temperature gradient or heat flow, slow and rapid thinning can be defined. As shown by the example in Figure 11, the mean lithosphere temperature (uplift) responds much more rapidly than surface heat flow to lithospheric thinning. For uplift purposes, thinning can generally be regarded as slow, whereas for surface heat flow considerations thinning is rapid, and the surface heat flow continues to increase after thinning stops.

The mode of the heat transfer though the base of the lithosphere during thermal thinning is problematical. If the lithosphere is stationary or only slowly moving (< 10 mm yr^{-1}) over the heat source, reasonable rates of thinning can be achieved by an increase in the heat flux through the base of the lithosphere of up to 10 times the equilibrium basal heat flux (Wendlandt and Morgan, 1982; Spohn and Schubert, 1982, 1983). From steady state models of mantle plumes (Parmentier et al., 1975; Yuen and Schubert, 1976), this increase in heat flux is reasonable. Turcotte and Emerman (1983), however, from stagnation point theory, conclude that thermal thinning would be too slow a thinning mechanism. If the plate is moving over the heat source, a dramatic increase in heat flux at the base of the lithosphere is required for significant thermal thinning (Birch, 1975; Gass *et al.*, 1978; Pollack *et al.*, 1981; Meng and Morgan, 1982). For example, Meng and Morgan (*op cit*) calculated that an increase in heat flux into the base of the lithosphere equivalent to a gradient on the order of 750 $^{o}C\ km^{-1}$ is required to thin the Pacific plate moving at 90 mm yr^{-1} from 90 to 45 km to produce the uplift and thermal anomalies of the Hawaiian Swell. Such a high apparent gradient indicates that the heat transfer is not simple conduction, although purely conductive models of the lithosphere appear to reasonably reproduce the thermal response of the Pacific plate to the thermal perturbation which produces the Hawaiian Swell (Von Herzen *et al.*, 1982; Meng and Morgan, *op cit*). The anomalous conductive heat loss from the Hawaiian Swell is approximately an order of magnitude greater than the volcanic heat loss from the Hawaiian Islands (Morgan and Phillips, 1983), which indicates that even if the heat enters the lithosphere nonconductively in this system, most of the heat leaves by conduction.

Although heat may enter the base of the lithosphere by some nonconductive process, magmatism, lithospheric delamination, or stoping of the lithosphere by the asthenosphere, the main heat transfer process within the lithosphere appears to be conduction. In this respect, as long as thinning takes place, the thermal form of the response of the lithosphere is the same for a stationary or moving lithosphere, i.e., the thermal response of a lithospheric column over an asthenosphere thermal anomaly is essentially the same as that for a stationary column with an equivalent basal heat flux history. Thus, although the mode of thermal thinning remains enigmatic, the thermal response of the lithosphere can be reasonably modelled by conductive models.

In the other end member of rifting, passive rifting, it is assumed that the lithosphere thins, or tends to thin, during lateral extension in response to a regional extensional stress field. The asthenosphere, or magmatic products from the asthenosphere, ascend to maintain volume as lithospheric material moves laterally. Heat is convected upward as there is a net rise in both the lithosphere and the asthenosphere during extension, and this has been suggested as a possible cause for high heat flow in rift zones (Osmaston, 1971; Morgan, 1973, pp. 213-214; Lachenbruch and Sass, 1978; McKenzie, 1978). In the simplest quantitative model of this process, the lithosphere is assumed to stretch uniformly and instantaneously, and surface heat flux increases instantaneously by a factor equal to the lateral extension factor (McKenzie, op cit). The stretching event is followed by a cooling and thickening of the lithosphere, and consequent decay in the surface heat flow anomaly. Jarvis and McKenzie (1980) have shown that for most practical purposes extension can be considered to be instantaneous if the stretching event lasts less than 20 m.y. In order to explain higher surface heat flow than predicted by the simple stretching model, the model has been modified to include variations in the amount of extension with depth (usually greater extension at depth), and stretching which includes a component of extension due to dyke intrusion (Royden and Keen, 1980; Royden et al., 1980). Both of these modifications cause a further increase of ths shallow geothermal gradient.

Lachenbruch and Sass (1978) have calculated the thermal effects on the lithosphere of continuous lithospheric extension, with the assumption that the lithospheric thickness is maintained constant either by crystal accretion onto the base of the lithosphere, underplating by basaltic magma, or magmatic intrusion. The predictions of the change in surface heat flow as a function of extension rate for these models is shown in Figure 12. As with the modified instantaneous stretching models, these continuous stretching models predict that the geothermal gradient is highest near the surface, and decreases with depth. When extension ceases, the lithosphere cools and thickens after a period of continuous extension in a similar manner to that predicted after instantaneous stretching.

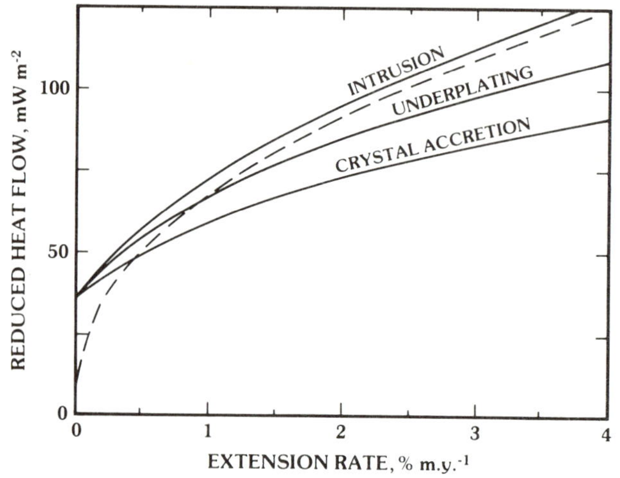

Figure 12. Variation in reduced heat flow as a function of constant extension rate for the crystal accretion, underplating by basaltic magma, and magmatic intrusion models of Lachenbruch and Sass (1978). Solid curves assume heat flux of 33 mW m^{-2} from asthenosphere. Dashed curve (intrusion) assumes 7.5 mW m^{-2} heat flux from asthenosphere. (Modified from Lachenbruch and Sass, op cit).

The purpose of the present discussion is not to argue the merits of active and passive rifting models: Different models may apply to different extensional events, and in most situations a combination of the models is probably applicable. Both mechanisms cause a similar thermal perturbation to the lithsophere; i.e., a thinning of the lithosphere and an increase of its thermal gradient. The magnitude of these effects may vary, and details of the thermal disturbance may change according to the importance of magmatic heat transport in individual cases. After the short term transient thermal disturbances due to isolated intrusions have decayed, however, the form of the basic themal perturbation will be the same, a high geothermal gradient in a thinned lithosphere.

B. Strike-Slip Tectonics

Major strike-slip faults in the continental lithosphere are associated with transform motions between lithospheric plates. In general, plate interactions in continental lithosphere occur over more diffuse zones than in oceanic lithosphere, and these strike-slip motions and thermal interactions can be complicated by movement along branching en echelon fault systems. Portions of transform boundaries where major strike-slip motion is offset may be zones of local lithospheric extension or compression. The discussion in this section is focussed upon thermal interactions during purely strike slip motion. Three types of thermal interaction can be associated with these faults: (a) frictional heating on the faults; (b) convective heat transfer by groundwater and/or magma in the fault zones; and (c) juxtaposition of lithospheric sections with different thermal structures across the faults.

Frictional heating on faults is strongly dependent upon the magnitude of the shear stress, tor, acting on the faults. If the two sides of a fault are moving with a uniform velocity of 2u relative to each other, the average rate of mechanical heat production, Q_{tor}, is given by

$$Q_{tor} = 2utor \qquad (9)$$

(e.g., Scholz _et al_., 1979; Lachenbruch and Sass, 1980). For significant heat production by this mechanism at typical plate velocities, the shear stress must be large: e.g., if the relative motion across the fault is 1 cm/yr, a 32 MPa (320 bars) stress is required for a frictional heat generation of 10 mW m^{-2}. Laboratory studies of rock friction generally predict that shear stresses on earthquake faults must be at least as high as 100 MPa, unless pore fluid pressures are very high, but seismological studies indicate that the average stress drop in crustal earthquakes is usually on the order of 10 MPa, or less (e.g., Atkinson, 1982). The thermal effects associated with strike-slip faulting should be a direct indication of the magnitude of the shear stress acting on fault zones, but seemingly contradictory effects have been reported from different fault zones.

Scholz _et al_. (1979) report on the effects of frictional heating on the Alpine fault of New Zealand, as preserved in the metamorphic record and thermally reset K-Ar and Rb-Sr ages of the rock units adjacent to the fault. Uplift of 4-11 km on the southeast side of this fault has exposed a record of metamorphic zoning subparallel to the fault for 350 km along strike, increasing in metamorphic grade from prehnite-pumpellyite facies 9-12 km from the fault, through the chlorite and biotite zones of greenschist facies, to the garnet-oligoclase zone of amphibolite facies within 4 km of the fault. The K-Ar and Rb-Sr ages of the rocks adjacent to the fault increase in age from 4 m.y. within 3 km of the fault to approximately 120 m.y. 20 km from the fault. Although the timing of movement along the Alpine fault is not rigorously constrained, Scholz et al. (op cit) argue that models of the thermal event or events recorded by the metamorphic zoning and radiometric age resetting require that the frictional shear stress acting on the fault during motion was at least 100-150 MPa.

In contrast, Brune et al.(1969) and Lachenbruch and Sass (1980) recently compiled heat flow data from the San Andreas fault zone, California, and conclude that there is no evidence for frictional heating of the main fault trace over a 1000 km length of the fault. Lachenbruch and Sass (op cit) consider it unlikely that heat transfer by hydrologic circulation masks the surface anomaly due to frictional heating, and with the resolution of the data conclude that the average shear stress on the fault does not exceed 10 MPa.

The apparently contradictory results from the Alpine and San Andreas faults can be reconciled if it is accepted that both high shear stress (100 MPa) and low shear stress (10 MPa) faulting occurs in the upper crust, the latter being more common (Sibson, 1980; Atkinson, 1982). High shear stresses are associated with dry fault zones, and there is evidence that the Alpine fault is exceptionally dry (Sibson et al., 1979). High pore fluid pressures can cause low stress faulting, although as discussed by Lachenbruch and Sass (1980), there are a number of factors that could contribute to the low average shear stress on the fault required by the San Andreas heat flow data. Alternatively, it can be argued that the metamorphic zonation associated with the Alpine fault represents primarily a vertical rather than a horizontal section of the crust, exposed by differential erosion and uplift, and that stresses on both the Alpine and San Andreas fault systems are similarly low.

With respect to the thermal structure of the continental lithosphere, frictional heating on strike slip faults must be considered only of local importance. Model results presented by Scholz et al. (1979) and Lachenbruch and Sass (op cit) show that the lateral effects of the heating are localized in the fault zone and are insignificant at distances greater than a few times the vertical extent of the fault, which is essentially limited by the thickness of the brittle layer of fracture in the lithosphere (on the order of 10 to 20 km). For high stress faults these effects can be locally very significant, with thermal anomalies of several hundred degrees in the upper crust. For low stress faults (the majority?), however, the effects will be minor.

Vertical faults can incorporate zones of high vertical permeability, and vertical fluid movements in these zones can generate thermal anomalies (e.g., Sorey and Lewis, 1976). Eckstein and Simmons (1978) and Eckstein (1979) present heat flow data from the Dead Sea fault zone which indicate significant shallow thermal anomalies associated with groundwater circulation partially controlled by increased permeability in this transform fault zone. In addition, although there is no direct evidence that major continental strike-slip faults act as conduits for magmatic activity, volcanism can be spatially associated with these fault zones (e.g., Eckstein, op cit, Figure 1), presumably with associated transient thermal anomalies. Neither groundwater circulation nor magmatic activity appear to be universal features of continental transform faults, however, and where they occur they can be treated as local anomalies rather than characteristic features of strike-slip plate boundaries.

Potentially the most dramatic thermal effect of a continental transform fault is the juxtaposition of lithospheres with different thermal structures by the fault.

This situation is common in oceans where lithospheres of different ages and thermal structures are juxtaposed across transform faults and fracture zones (e.g., Sibuet and Mascle, 1978). The growth of the San Andreas fault, by the north-northwestward migration of the Mendicino triple junction relative to the North American plate, changes the thermal and tectonic setting of the eastern margin of the North American plate, from overriding the subducting Gorda plate north of the triple junction, to abutting the Pacific plate south of the junction (see Lachenbruch and Sass, 1980; Zandt and Furlong, 1982). This transition is marked by a heat flow increase by a factor of 2 over a broad zone in the Coast Ranges of California within 4 m.y. after the passage of the triple junction (Lachenbruch and Sass, op cit).

The thermal anomaly associated with the growth of the San Andreas fault is thought to be related, at least in part, to the replacement of the cold, subducting Gorda plate beneath the margin of the North American plate by hot asthenosphere after the passage of triple junction, and thermal models of this transition have been presented by Lachenbruch and Sass (1980) and Zandt and Furlong (1982). To explain the heat flow data, these models require that the lithosphere of the eastern margin of the North American plate is thinned to around 20 km after the passage of the Mendicino triple junction, and the thermal effects on the lithosphere are not unlike the passage of the plate over a hot spot (e.g., Von Herzen et al., 1982; Meng and Morgan, 1982). Zandt and Furlong (op cit) argue that thinning to 20-30 km is not unreasonable, and present seismic data to support their arguments. Lachenbruch and Sass (op cit) suggest that thinning may not be quite so drastic if shear heating at the base of the seismogenic layer contributes to the surface anomaly, and although heating from the shallow asthenosphere must be a significant factor in the anomaly, the relative contributions of the two mechanisms are at present unresolved.

In summary, strike-slip or transform faults in continental lithosphere can have a variety of thermal effects on the thermal structures of the lithosphere adjacent to the fault. If the shear stress acting on the fault is high (100 MPa), significant frictional heating can occur on the fault at shallow (upper to mid crustal) depths. Most faults appear to have low shear stress (10 MPa), however, in which case, unless relative motion across the fault is very fast (tens of cm yr^{-1}), frictional heating will be minor. Groundwater circulation and magmatic activity associated with the faults may be of local thermal importance, but are not universal features of transform faults. Where a transform fault changes the nature of a plate boundary, as in triple point migration and growth of the fault, major changes can occur in the thermal structure of the lithosphere at the plate boundary in response to the new boundary conditions.

C. Zones of Convergence

According to the natures of the two plates involved in the interaction, zones of convergence can be classified as island arc (ocean/ocean), continental arc (continent/ocean) or continental collision (continent/continent). In many respects the island arc and continental arc are similar in that they both involve the subduction of oceanic lithosphere, but for the thermal structure of the continental lithosphere we are primarily concerned with continental arc and continental collision plate convergence. Island arc systems are important to the continents when they are accreted to the continents in arc-continent collisions, but this can be regarded as a type of continental collision. Of the two continental types convergence zones, continental arcs are structurally simpler, and data from several such zones on the margin of the Pacific give a relatively consistent picture of at least the shallow thermal structure of the lithosphere in these zones.

A thermal characteristic of continental arcs is the development of paired metamorphic belts, blueschist on the oceanic side, and high temperature on the continental side (e.g., Dewey and Bird, 1970, Cordilleran-type mountain belts). These belts indicate low temperatures in the lithosphere of the continental margin, with higher temperatures inland. Heat flow data from the Japan arc (Hasebe *et al.*, 1970; Watanabe *et al.*, 1977; Anderson, 1980), the Sunda arc (Carvalho *et al.*, 1980), the central Andes (Henry, 1981; Uyeda and Watanabe, 1982), and the North American Cordillera (Hyndman, 1976; Blackwell *et al.*, 1982) confirm this basic thermal pattern. In all of these data sets the inland heat flow high is evident, and all show the "coastal" low except the Sunda data set for which no measurements are available in this zone.

Significantly, however, not all areas in the Andes show the inland high heat flow, and Henry (1981) reports low inland heat flow (20-40 mW m^{-2}) for the northern central Andes. Henry (*op cit*) makes a correlation between low inland heat flow and low angle subduction, suggesting that the area of low heat flow may overlie a zone where the asthenosphere does not penetrate between the converging plates, a concept supported by low values of seismic attenuation in the upper mantle in this area, and strong coupling between the plates causing large earthquakes. To the south, high inland heat flow (80-100 mW m^{-2}) occurs in a zone with a higher subduction angle, high upper mantle seismic attenuation, and a minimum of seismic energy release. A transitional zone with intermediate inland heat flow (40-80 mW m^{-2}) separates the high and low heat flow zones, and the heat flow is thought to decrease again to the south where the subduction angle again becomes shallower (Henry, *op cit*).

The transition between the low coastal and high inland heat flow zones has been constrained to be very sharp, approximately 20 km, in part of the Oregon section of the North American Cordillera (Blackwell *et al.*, 1982). Similar sharp transitions are indicated farther north in the Cordillera in British Columbia (Hyndman, 1976), and in the central Andes (Henry, 1981). Anderson (1980), however, reports a more gentle transition in the fore arc basin of the Japan arc and other southeast Asian arcs, with normal heat flow values in front of the volcanic arc. In the central Andes and the British Columbia Cordillera the heat flow transition coincides with the first occurrence of recent volcanic centers (the volcanic arc; Henry, op cit; Hyndman, op cit), although in the Oregon Cordillera the active volcanoes are approximately 20-30 km inland of the transition (Blackwell et al., op cit). The transitions in both Oregon and British Columbia coincide with the occurence of hot springs and the transition inland from high to low Bouguer gravity (Blackwell et al., op cit), and also in British Columbia with a transition inland to low deep crustal and upper mantle electrical resistivity, and recent regional uplift (Hyndman, op cit).

Several authors have concluded that high heat flow in arcs is related to magmatic activity (e.g., Oxburgh and Turcotte, 1971; Hyndman, 1976; Henry, 1981; Blackwell et al., 1982). From the widths of the Oregon heat flow and gravity transition zones, Blackwell et al. (op cit) postulate the presence of a hot, low density region, about 60 km wide, in the crust below 7-10 km, probably a zone of temporary residence of magmas melting off the subduction zone. This general concept would appear consistent with data from all the continental arcs with high heat flow regions. Thus high thermal gradients are clearly maintained in the upper crust by magmatic activity, but this activity masks the thermal processes in the lower crust and upper mantle over the subduction zone.

All available heat flow data sets for continental arcs indicate low to normal heat flow, on the order of 40 mW m^{-2}, in the "coastal" zone between the trench and the volcanic arc. Thermal models of subduction zones show that low heat flow is to be expected close to the trench due to downward convection of heat by the subducting slab (Minear and Toksöz, 1970a,b; Toksöz et al., 1971; Hsui and Toksöz,1979), and the penetration of low temperatures to depth is recorded in the blueschist (high pressure, low temperature) metamorphism developed in the outer belt of active margins. Data from the northern region of the central Andes studied by Henry (1981) show that the low heat flow zone (20-40 mW m^{-2}) penetrates inland beyond the normal "coastal" zone, into an area of mid-Miocene arc volcanism. Assuming the volcanism to indicate high heat flow typical of other inland active margin regions during the volcanism, the data indicate a rapid decrease in heat flow during the past 10-15 m.y A similar heat flow decrease in part of the Oregon Cordillera is indicated by data presented by Blackwell *et al.* (1982): A 40-60 km westward migration of the volcanic arc from its position in the Oligocene and Miocene, related to a shallowing of the angle of subduction, has been accompanied by a migration of the low/high heat flow transition, and in the wake of this migration heat flow has decreased from high inland values to low coastal values during the last 6-10 m.y These rapid decreases in heat flow provide constraints on the thermal processes associated with subduction.

Two basic sources for subduction related magmatism have been proposed; the upper surface of the subducting plate, and the overlying mantle (e.g., see Marsh, 1979). As subduction is primarily a cooling process (Hsui and Toksöz, 1979), whatever source region is appropriate, either a lowering of the melting temperature, or an input of heat must take place to generate magmas. Perhaps melting might be induced in the mantle overlying the subducting slab by the influx of fluids from the sediments of the slab, but Marsh (*op cit*) argues that this mechanism is inapplicable as the volcanic products generated by this mechanism would be compositionally different from those observed, and magmas generated by this mechanism would probably solidify before reaching the surface. Alternatively, numerous authors have suggested that heat is input into the system by viscous heating on the upper surface of the descending slab, and by flow in the asthenosphere induced by the motion of the descending slab (e.g., Turcotte and Schubert, 1973; Marsh, 1976, *op cit*; Hsui and Toksöz, *op cit*). Under certain conditions, viscous heating may lead to a thermal feedback instability and magma generation (e.g., Spohn, 1980).

However, the rapid decreases in heat flow described above, associated with a shallowing of the angle of subduction and the migration or cessation of arc volcanism, suggest that flow in the asthenosphere, cut off by a shallowing of the subducting plate, is more important than viscous heating, which should continue, and perhaps increase with shallower subduction.

The relatively short time constant, 6-10 m.y. for the Oregon Cordillera and 10-15 m.y. for the central Andes, indicates that the source of the high heat flow is shallow. The maximum depth of the source can be estimated from the characteristic thermal time constant for a slab, $t_{char} = (L^2/4k)$, where L is the slab thickness and k is thermal diffusivity. Assuming a thermal diffusivity of 32 $km^2/m.y.$, L is 28 km for $t_{char} = 6$ m.y., and the source of the thermal anomaly is constrained to be at or shallower than this depth. This result is consistent with the shallow depth of intense thermal disturbance of 10 ± 2 km postulated by Blackwell *et al.* (1982) from the width of the low/high heat flow transition. Although some heating must also occur below this depth, temperatures significantly above the postulated intrusion temperature ($700 \pm 100\,^{o}C$; Blackwell *et al.*, *op cit*) do not appear to occur above depths on the order of 25 km, or higher heat flow should still be measured in the wake of the migrating volcanic arc. Thus the thermal structure of the high heat flow lithosphere overlying a subducting plate would appear to be very high gradients down to the depth of major intrusion, below which the gradients are normal to low. The development of this type of thermal profile by repeated intrusions at relatively shallow depth has recently been discussed by Wells (1980).

Back-arc extension is related to subduction, and is observed behind both island arcs (e.g., the Lau Basin behind the Tonga arc system) and continental arcs (e.g., extension in the Altiplano of Peru, and the mid-Cenozoic North American Basin and Range extension). During initial stages of development of these areas, the thermal structure of their lithosphere is probably close to that described above for the high heat flow region of the continental arc. As extension continues, however, extensive lithospheric thinning must take place, and there is a transition to the thermal regimes described under extensional tectonics above. Any stage in between the two thermal regimes is possible depending upon the ratio of subduction magmatism to extensional heating.

The thermal regime of continental collision zones is very complex due to the variety of heat transfer processes that can occur in these zones. Surface heat flow generally appears to be high in these zones, but varies rapidly across strike, and is highly variable due to the complex shallow thermal regime in these areas of high relief, complex structure and rapid erosion (see e.g., Cermak and Hurtig, 1979; Cermak, 1979). Surface heat flow values in the Alps are typically in the range 70-80 mW m^{-2} (Cermak, op cit) but much, if not all of this high heat flow has been ascribed to the effects of erosion (Clark and Jäger, 1969; Haenel, 1974; England and Richardson, 1980).

Unfortunately, detailed heat flow determinations are not available for the zones of continental collision east of the Alps. Coster (1947) gives a mean heat flow of 36 mW m^{-2} for the Masjid-i-Sulaiman oilfield in the Zagros convergence zone, and Bird *et al.* (1975) have used oil well temperature data to estimate values increasing from 38 to 48 mW m^{-2} across the zone of underthrusting towards the Persian Gulf. Thus, this convergence zone appears to be characterized by low to normal heat flow. No heat flow data are available at present from the Himalayan collision zone, but high heat flow is indicated in the elevated regions of the Himalaya and Tibet Plateau by abundant hydrothermal activity and Quaternary volcanic cones (Wang *et al.*, 1982; Zhang *et al.*, 1982). Unlike the data from continental arcs, the continental collision heat flow data cannot be used directly to constrain the thermal structure and processes in these zones, but can only be used as boundary conditions for thermal models.

Thermal and mechanical models of the crust and upper mantle in continent/continent convergence zones show that the thermal structure of the resulting lithosphere changes both across strike, and in time as the collision zone develops (Bird et al., 1975; Toksöz and Bird, 1977; Bird, 1978a,b). The sequence of events postulated to have occurred during collision in the Himalayan-Tibetan region is shown in Figure 13. Prior to continental collision subduction of the intervening oceanic plate created a continental arc on the overriding continent, similar to the examples decribed above (Figure 13A). Some magmatic thickening of the overriding continental crust may have occurred at this time; the amount of heating of the overriding plate is related to the duration of subduction, and may be important in subsequent deformation of the overriding plate (Figure 13B). The closing of the intervening ocean was marked by the subduction of continental rise sediments, then underthrusting of the continental shelf of the continent of the subduction plate. The underthrust plane became the suture zone between the two continents, and at this stage further subduction of the underthrusting plate was prevented by the buoyancy of its continental crust. Continued convergence between the two continents was accomodated by deformation in the heated continental arc of the overriding plate (Figure 13C).

After the formation of the continental suture in the Himalayas, granites where formed in the underthrusting continent in a zone on the order of 100 km wide, to the south of the suture. Bird (1978a) calculated that these granites could not have been formed by the frictional heating of thrusting, and postulates that the mantle portion of the lithosphere delaminated from the underthrusting plate, exposing the base of the underthrusting continent to asthenospheric temperatures, and inducing melting (Figure 13C). Increased resistance to convergence eventually led to the development of a second zone of underthrusting in the underthrusting plate (Figure 13D), and a double section of continental crust was formed. Continued convergence was

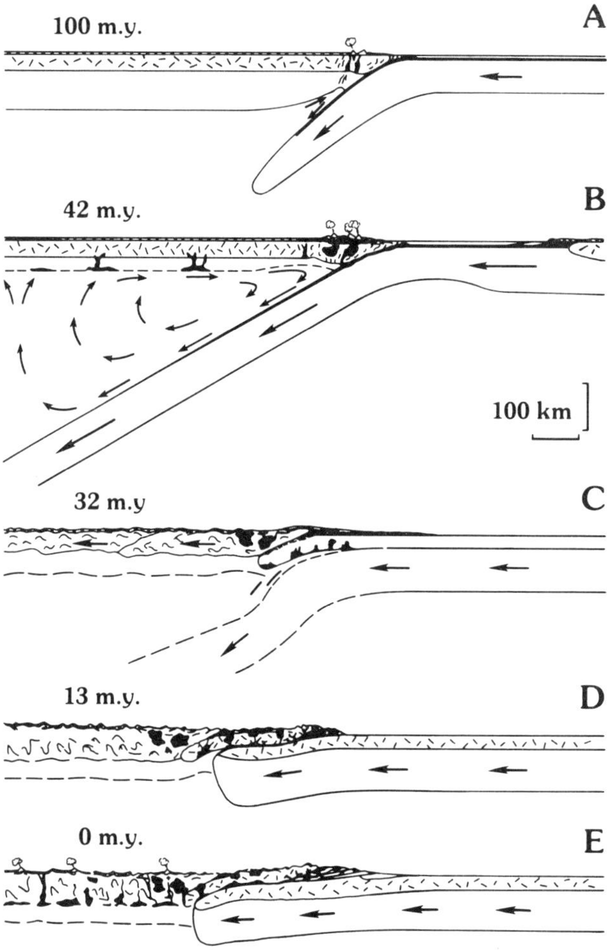

Figure 13. Tectonic development of the Himalayan-Tibet region. (After Toksöz and Bird, 1977).

again accomodated by deformation in the original overthrusting plate (Figure 13E), which was accompanied by volcanism in the Tibetan Plateau (Toksöz and Bird, 1977; Wang *et al.*, 1982).

The results of thermal models by Toksöz and Bird (1977) of three stages of continental collision are shown in Figure 14: A) after underthrusting of the continental margin; B) after crustal doubling; and C) 35 m.y. after the cessation of convergence. The surface heat flow profile predicted by this type of model for the Zagros collision zone matches the limited thermal data for this area (Bird *et al.*, 1975). Unfortunately the data do not constrain the models well, but the models are basically consistent with

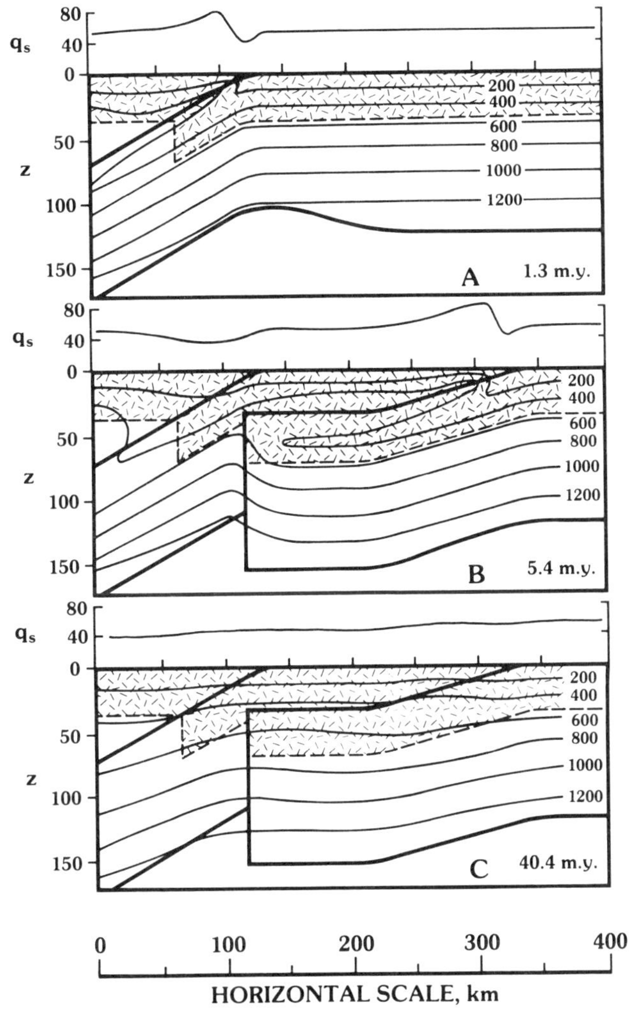

Figure 14. Thermal models of three stages in continental collision. For 1.3 m.y. the edge of the continental crust (stippled) subducts at 50 mm $year^{-1}$ with the oceanic plate (A). Buoyancy then prevents further subduction, and underthrusting begins at the right, proceeding for 4.1 m.y. until the crust is doubled (B). The calculation then proceeds for 35 m.y. without further motion (C). Depth z is given in km. Temperatures are given on isotherms in ^{o}C, and surface heat flow q_s is shown above each section in mW m^{-2}. (After Toksöz and Bird, 1977).

the thermal and structural phenomena of continental collision, and allow the following generalizations to be made about the thermal stucture of the lithsophere during continental collision. Prior to collision the overriding plate is heated by magmatism over the subducting oceanic plate (Figures 13A and 14A). The duration of this heating may control the extent of later deformation in the overriding plate, ranging from minor with the subduction of less than 500 km of oceanic crust in the Alps, to major with the subduction of some 8000 km of oceanic plate under the Tibetan Plateau (Toksöz and Bird, op cit). During the time period between formation of the continental suture and doubling of the underthrusting plate, some heating of the underthrusting plate is indicated by granite formation in this plate as discussed above. The doubling of the underthrusting plate is accompanied by a net depression of the geotherms in the resulting lithosphere, and temperature inversions may also occur, as discussed below (Figure 14B). After the cessation of convergence, geotherms in the lithosphere will relax to steady state conductive geotherms (Figure 14C).

In addition to the two dimensional thermal modelling described above, several authors have made more detailed one dimensional thermal models of the crust where it is doubled in thickness by underthrusting (e.g., Oxburgh and Turcotte, 1974; Bickle et al., 1975; Graham and England, 1976; England and Richardson, 1977; Richardson and England, 1979). These models basically study the relaxation of a "sawtooth" geothermal profile formed by the superposition of the undisturbed gradients in the overthrust and underthrust crust, and are used to explain metamorphic facies exposed by erosion in collision zones. Viscous heating in the thrust zone is commonly included in these models (e.g., Bickle et al., op cit). By including the effects of erosion into the models, the temperature history of rock units brought to the surface can be calculated (England and Richardson, op cit; Richardson and England, op cit). Although many parameters in these models are poorly constrained, such as the erosional history and mantle heat flow, and the models do not include convective heat transport by metasomatic fluids, the conditions of metamorphism can usually be met with "reasonable" parameter values in the models. Thus, the thermal history of the overthrust zone is described reasonably well by the relaxation of an overturned geotherm, possibly with additional heat input by magmatism and/or high mantle heat flow, but the details and relative importance of different thermal processes in these zones are poorly constrained.

The net effect of overthrusting is to depress the geotherms in the resulting lithosphere, although overthrust crust may be hot prior to overthrusting (Bird, 1978a). Magmatism ("post-tectonic granites") is not uncommon in continental collision zones after the cessation of convergence, however, and thus some reheating of the lithosphere appears to occur. This reheating could be simply thermal relaxation of the depressed geotherms, with a modest component of mantle heat flow causing melting at the base of the thickened crust. By this mechanism the collision zone would heat up to the stable continental geotherm. Alternatively, Houseman et al. (1981) have suggested that the mantle portion of the lithosphere, thickened during underthrusting, may be mechanically unstable and sink, being replaced by hot asthenospheric material which induces magmatism. The mantle lithosphere detachment would result in a thin, hot lithosphere, which would subsequently thicken and cool to a stable continental geotherm.

Heating or cooling of the thickened continental crust after collision is masked at the surface by the upward convection of heat as the lithosphere rises by the effects of isostatic rebound as material is removed from its surface by erosion. This process would speed the relaxation of the depressed geotherms, and may even cause relative elevation of the geotherms in the upper lithosphere. England and Richardson (1980) have suggested that the general decay in surface heat flow with time since the last tectonothermal event mainly reflects the effects of erosional processes, and should not be used to infer directly the thermal development of the subcrustal lithosphere (see below). While this observation may be correct in continental collision zones, however, it is inappropriate in areas of extensional or hot spot tectonics, where erosion is minor.

D. Uplift, Downwarping, Erosion, and Sedimentation

Uplift, downwarping, erosion, and sedimentation all change the surface temperature boundary condition of the lithosphere and therefore cause a thermal perturbation within the lithosphere. In many instances, these effects are minor, and can be considered as shallow thermal processes. After major tectonic activity, however, crustal thickening may initiate a long period of erosion with isostatic uplift, or crustal thinning may cause a long period of sedimentation with isostatic downwarp. In these instances, the thermal perturbation to the lithosphere, especially to the surface heat flow, may be relatively long lived. The problem can be further complicated by thermal relaxation in the lithosphere after a tectonic event, and the addition or removal of heat producing material.

England and Richardson (1980) have suggested that the effects of erosion on surface heat flow can be significant for a time period up to four times the duration of erosion. If upper crustal radioactivity is ignored, the surface heat flow increases until the end of erosion, and then decreases. If crustal heat production is included in the analysis, much less increase in surface heat flow is predicted during erosion, and the heat flow begins to decrease prior to the end of erosion due to the removal of radioactive material from the crust. Examples of these effects are shown in Figure 15. England and Richardson do not discuss the lithospheric thickness used in these calculations, and from the long time decay of the thermal anomaly after erosion, an effectively infinite thickness lithosphere is implied by their results. If a limited thickness lithosphere was used, the decay of heat flow after erosion would probably be slightly faster, more on the order of the relaxation times discussed below. Also England and Richardson (op cit) did not include initial thermal perturbations in the lithosphere thermal structure in their models. The inclusion of these effects would cause additional changes to the surface heat flow during and after erosion. A more complete analysis of the thermal effects of tectonism and erosion is given by

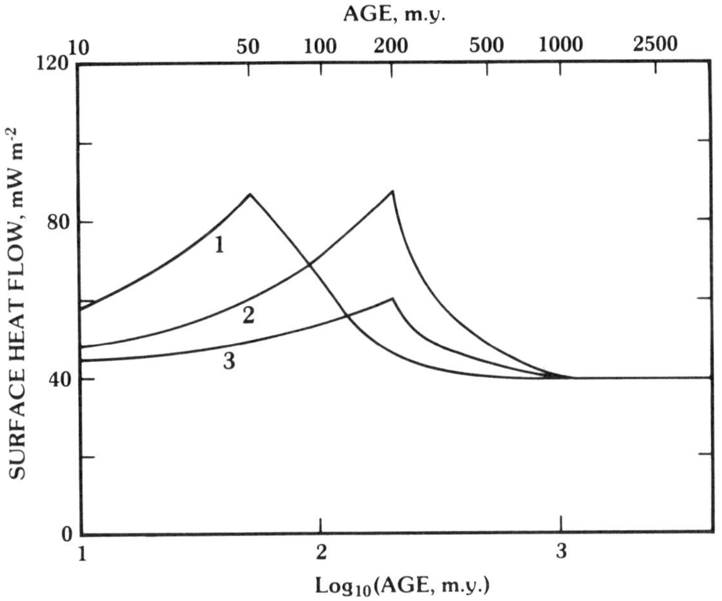

Figure 15. Surface heat flow as a function of time since the start of erosion for different depths of erosion, Emax, and erosion periods te, assuming that erosion depth E(t) at time t is given by

$$E(t) = (t/te)Emax \qquad t < te$$

$$E(t) = Emax \qquad t \geq te$$

A uniform undisturbed heat flux of 40 mW m^{-2} is assumed. For 1) Emax = 30 km, te = 50 m.y.; 2) Emax = 60 km, te = 200 m.y.; and 3) Emax = 30 km, te = 200 m.y. (After England and Richardson, 1980).

Kono and Amano (1978), who show that high heat flow immediately following tectonism can mask the initial heat flow increase during erosion predicted by England and Richardson (op cit). As the thermal state of the lithosphere after tectonism is poorly defined, however, it is impossible to usefully evaluate the short term thermal implications of these models. It is obvious, however, that in continental convergence zones where significant crustal thickening occurs, erosion after tectonism can alter and prolong the thermal relaxation of the lithosphere.

The effects of sedimentation are opposite to those of erosion (see Benfield, 1949; Jaeger, 1965). The addition of sediment depresses the near surface geothermal gradient and heat flow, while the addition of heat producing elements within the sediment tends to increase the heat flow. As with erosion, these effects will be greatest during sedimentation, but there will also be a period of thermal relaxation after the cessation of sedimentation. The details of the thermal processes during sedimentation are commonly more complex than those during erosion, as young sediments will undergo diagenesis and compaction during burial, and groundwater flow through porous sediments can cause significant nonconductive heat transfer. Recently thermal models have been used to predict the thermal evolution of sedimentary basins, and the thermal histories of the sediments within the basins (e.g., McKenzie, 1978; Royden and Keen, 1980; Royden et al., 1980). However, although these models may reasonably represent the thermal evolution on a lithospheric scale, the complexities of the thermal properties and thermal processes within the basin sediments limits the use of the models for accurately predicting the thermal histories of the sediments. In general, sedimentation does depress the surface heat flow in a sedimentary basin, although because of water flow problems within the basins, these sites are commonly avoided for heat flow determinations, and reliable thermal data from sedimentary basins are relatively rare.

E. Thermal Relaxation of the Lithosphere Following Tectonism

The tectonic processes discussed above all generate thermal perturbations in the lithosphere. As tectonism wanes, anomalous temperatures within the lithosphere will relax towards equilibrium values, possibly with some readjustment in lithospheric thickness. As most tectonic processes cause heating in the lithosphere, tectonism will usually be followed by cooling, and possibly thickening of the lithosphere. In convergence zones geotherms may be depressed, and if the lithosphere is thickened beyond its equilibrium thickness, it will thin after tectonism through input of heat from the asthenosphere (see thermal thinning above). The processes of thermal relaxation of the lithosphere are discussed in this section: observations concerning the equilibrium thermal structure and thickness of the continental lithosphere are considered in the following section.

There are two basic modes in which the continental lithosphere may thermally relax after tectonism; thermal conduction within the lithosphere, and readjustment of the lithospheric thickness. Relatively small and/or shallow perturbations, such as crustal plutons, will cool essentially conductively (e.g., see Jaeger, 1968), with cooling assisted at shallow depths by hydrothermal convection (e.g., Norton, 1977; Norton and Knapp, 1977; Norton and Knight, 1977), but these perturbations are short lived on a geological time scale (e.g., Harder _et al_., 1980).

Major thermal perturbations within the lithosphere have been modelled by Vitorello and Pollack (1980) by the conductive cooling of a triangular form temperature disturbance which disappears at the surface and at the base of the lithosphere. The upper boundary condition for this problem is well defined: ground surface temperature is essentially independent of heat flow, and thus the surface is maintained at arbitrary zero. The lower boundary condition is more difficult to define, however, and three lower boundary conditions were considered by Vitorello and Pollack (_op cit_): 1) temperature disturbance at the base of the lithosphere remains zero; 2) the lithosphere and asthenosphere are a continuous conductive medium, i.e., the problem is that of an initial triangular shaped temperature disturbance at the top of a semi-infinite medium; and 3) no heat flux across the lithosphere base. Examples of the results of the models with these three boundary conditions are shown in Figure 16.

The first boundary condition of Vitorello and Pollack (1980) maintains the condition that the base of the lithosphere remains at a constant temperature, which is reasonable if the lithosphere/asthenosphere transition is a rheological boundary. However, the only way in which this boundary condition can be maintained at a fixed depth during cooling is if the temperature profile in the lithosphere is inverted below the base of the lithosphere, i.e., an equivalent negative temperature disturbance at the top of the asthenosphere. As this thermal state in the convecting asthenosphere is highly improbable, the first boundary condition must be considered as highly improbable. The second boundary condition gives results intermediate between the constant temperature and no heat flux boundary conditions (Figure 16), which indicate that the temperature at the base of the asthenosphere rises, and that heat is lost from the lower lithosphere to the asthenosphere. The temperature rise will be discussed below, but as with the constant temperature boundary condition, the loss of heat into the asthenosphere in the semi-infinite model implies higher temperatures in the lithosphere than in the underlying asthenosphere, again an improbable condition.

Vitorello and Pollack (1980) suggest that the no flux boundary condition can only practically be maintained if the temperature perturbation is mirrored across the lithosphere base, and thus, the effective lithosphere thickness should be twice the thickness used in the model. A second mechanism for maintaining this condition is suggested, however, which does not require a doubled lithosphere to maintain the boundary condition. If the assumption that the asthenosphere is convecting is accepted, then, the minimum gradient in the asthenosphere boundary layer with the lithosphere will be the adiabatic gradient. This gradient is positive downwards, and therefore there will be a net upward flow of heat into the lithosphere. Temperature perturbations in the lithosphere are restricted in magnitude to magmatic temperatures, and as these temperatures are unlikely to exceed the asthenosphere temperature, temperatures within the lithosphere are unlikely to exceed the temperature at the top of the asthenosphere. Thus, although negative temperature gradients may be set up locally in the lithosphere, they cannot be set up between the lithosphere and asthenosphere. An approximate condition of no heat flux into the base to the lithosphere will be met when the lower lithosphere is heated close to the asthenosphere temperature, and this heat flux would increase as the lithosphere cools. A strict interpretation of this

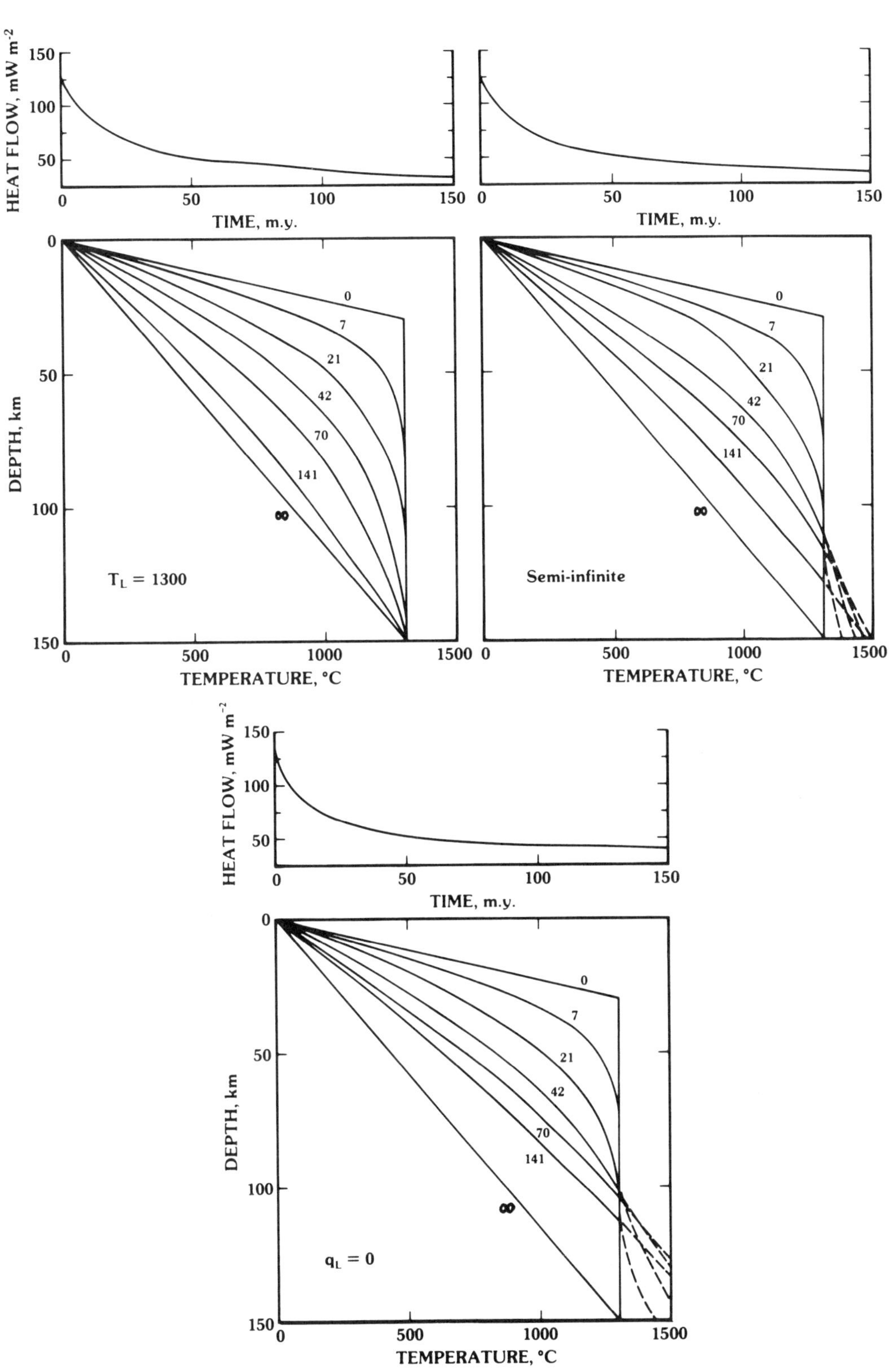

Figure 16. Conductive relaxation of a triangular form thermal perturbation in a fixed thickness lithosphere, assuming a steady-state geotherm equivalent to a 150 km thick lithosphere (see Figure 19), and a maximum temperature perturbation at a depth of 30 km. The three sets of curves are for three different lower boundary conditions: constant temperature, the lithosphere and asthenosphere are a continuous conductive medium, and no heat flux (see text). (Adapted from Vitorello and Pollack, 1980).

configuration would require a boundary condition of the heat flux into the base of the lithosphere $q_L \geq 0$, but the condition $q_L = 0$ should be a reasonable first approximation to this condition for heating events in the lithosphere Thus, in contrast to Vitorello and Pollack (1980) who preferred the first two lower boundary conditions, from arguments that consider convection in the asthenosphere, the no heat flux lower boundary condition is preferred.

If the no heat flux lower boundary condition is the most appropriate of Vitorello and Pollack's (1980) models for cooling of the continental lithosphere, then the temperature rise at the base of the lithosphere implied by this model must also be investigated. The rise in temperature at the base of the lithosphere predicted by the no flux model would result in a decrease in viscosity of the lower lithosphere, and from the definition of the thermal lithosphere used in this paper, this would imply conversion to asthenosphere, and thinning of the lithosphere. However, the low thermal gradient in the newly formed asthenosphere would probably prevent convection in this layer, and thus, cooling of the layer and the lithosphere above would be dominantly conductive, and follow the path predicted by the conductive cooling model.

An alternative, possibly complementary treatment of the relaxation of thermal perturbations in the lithosphere, is to assume that the lithosphere is thinned with or by the perturbation, and that relaxation involves cooling and thickening of the lithosphere. Solutions for an analytical approximation and a numerical solution for the problem of thermal thickening of the continental lithosphere have been given by Crough and Thompson (1976a,b), and these solutions suggest that the continental lithosphere cools in a similar manner to the oceanic lithosphere, the cooling and thickening curve for the continents being the tail end of the oceanic cooling curve. The predicted cooling time is somewhat dependent on the magnitude of the initial perturbation, but depends primarily on the equilibrium heat flux into the base of the lithosphere, the lower the basal heat flux, the longer the cooling time. With basal heat fluxes in the range 23-33 mW m^{-2} Crough and Thompson (1976b) predicted a 500-1000 m.y. time scale for the lithosphere to reach thermal equilibrium.

Kono and Amano (1978) performed a similar thickening analysis to Crough and Thompson (1976a,b), but also included the effects of latent heat in the transformation of asthenosphere to lithosphere during thickening. The addition of latent heat during thickening slows the thickening process. Kono and Amano (op cit) claim that the thickening rate is 10 times faster for an oceanic plate than for a continental plate, but at least part of this difference can be explained by the fact that continental lithosphere rarely, if ever, thins to the mid-ocean ridge lithospheric thickness, and thus, as suggested by Crough and Thompson (1976a), starts cooling at an advanced point on the cooling curve relative to the oceanic lithosphere. In addition, the inclusion of latent heat in the analysis implies significant partial melting at the base of the lithosphere. No such zone of partial melt has been detected by seismic methods under stable continental regions (e.g., Jordan, 1975), which implies that either the asthenosphere is absent under shields, or that latent heat may not be significant.

Before considering the consequences of cooling within the lithosphere and thermal thickening of the lithosphere, a note should be made that cessation of tectonism is not necessarily the time at which all thermal relaxation of the lithosphere begins. As discussed above, erosion or sedimentation can continue to perturb the lithosphere geotherm after tectonism, and will extend the time scale for thermal equilibrium to be attained. Interestingly, from cooling models of the oceanic lithosphere (e.g., Parsons and Sclater, 1977), very similar results are obtained from constant thickness lithosphere (plate) models and thermal thickening (cooling boundary layer) models if the maximum lithospheric thickness in the latter model is constrained to be the same as the constant plate thickness. With this same constraint, the conductive cooling lithosphere model, with the no heat flux lower boundary condition, and thickening lithosphere model discussed above yield similar results, if latent heat is ignored in the thickening model. This result is realistic because convection in the upper asthenosphere in the thickening model is suppressed by a decrease in thermal gradient caused by the perturbation above. Thus, if the lithosphere does have an equilibrium thickness, it is this thickness, and not the mode of cooling, that is the main factor in determining the time scale for equilibrium to be reached. With this result, it would be expected that the thermal perturbation following the cessation of erosion or sedimentation would decay on the same time scale as any other tectonothermal perturbation to the lithosphere (*cf* England and Richardson, 1980).

7. THERMAL EVOLUTION OF THE CONTINENTAL LITHOSPHERE

Surface heat flow integrates a number of components which can rarely be uniquely separated. These components include heat flow modification by near surface thermal regime, heat production in the continental crust, remnant heat from tectonism and magmatism, and an equilibrium heat flux component of sub-crustal heat flow. With the possible exception of the last of these components, they are all expected to be spatially and temporally variable, and examination of the specific data sets is required to isolate their contributions to the global heat flow data set.

A. Near Surface Thermal Regime

The near surface thermal regime mainly causes a redistribution of heat, thus tending to produce scatter in the surface heat flow data about a mean. Many of these processes, such as hydrothermal circulation, thermal refraction, topographic effects, sedimentaion and erosion, will be most important when the topographic relief is greatest and while structural relief juxtaposes poorly consolidated sediments with older rocks. Thus, there should be a trend of decreasing scatter in the surface heat flow data with increasing time since tectonism, and unless heat flow sites are selected with great care to minimize effects of the near surface regime, or elaborate corrections for topography and other effects are applied, it is generally true that data from young terrains show the largest scatter. If sufficient data are available, however, only systematic modification of the heat flow by the near surface thermal regime and bias introduced into the data by non-random sampling are important in studying the thermal evolution of the lithosphere.

Sampling bias may be introduced by the preference of basement rock sites for heat flow measurements, as these sites commonly have less obvious hydrothermal circulation than less consolidated rock sites. Heat flow may be systematically high in the basement sites due to thermal refraction in the higher conductivity basement (e.g., Harder et al., 1980) and erosion at these sites. Many heat flow measurements are made in mineral exploration boreholes, and thus it could be argued that the lithology at these sites is not regionally typical. In addition, there tends to be some unconscious filtering of data by investigators: zero or very low gradients are discarded as "obviously disturbed by hydrothermal circulation," but high gradients tend to be reported as "interesting." Quantitative assessment of the importance of these factors is impossible, but they probably tend to cause a minor increase in the apparent mean heat flow. Sampling problems can only be resolved by an increase in data coverage and density.

In addition to sampling problems, some features of the near surface thermal regime may make a systematic contribution to the near surface heat flow. The effects of erosion and sedimentation have already been discussed, and in addition to these effects, Quaternary climatic effects can cause a significant systematic perturbation to surface heat flow data (Birch, 1948), typically a decrease in the surface heat flow. For some data sets, such as the Superior Province of Canada, corrections for Pleistocene climatic variations seem justified (e.g., see Allis, 1978), and the corrected data seem more compatible with other data sets than the uncorrected data (e.g., see Jessop and Lewis, 1978). In other areas, such as the unglaciated eastern United States, the validity of climatic corrections is questionable. Thus, as with sampling problems, quantifying the net effect of climatic variations on the heat flow data set is difficult: mean heat flow is likely to be slightly underestimated without this correction, which may tend to offset the effect of sampling problems.

Except for the effects of sedimentation and erosion, which will be considered with the remnant heat of tectonism and magmatism below, the near surface thermal regime is responsible primarily for scatter in surface heat flow data, which may decrease with time after tectonic activity, with possibly a small, but unresovled, bias to the data.

B. Crustal Heat Production

In the discussion of steady-state conduction in the continental lithosphere above, the linear relationship between surface heat flow and surface heat production in plutonic and high grade metamorphic terrains, and the concept of heat flow provinces were discussed without regard to lateral or temporal variations in crustal heat production. On a global scale, Pollack and Chapman (1977) derived an empirical relationship between surface heat flow and reduced heat flow which indicates that approximately 40% of the surface heat flow come from the crustal zone enriched in heat producing elements. Crustal heat production is thus an important term in the thermal structure and evolution of the continental lithosphere. It has been suggested the erosion can remove a significant portion of the crustal heat producing isotopes (e.g., Kono and Amano, 1978; England and Richardson, 1980), and that this removal of crustal heat production is a significant factor in the apparent secular decrease in global heat flow (e.g., Vitorello and Pollack, 1980). Little discussion has been given to the redistribution of heat producing elements during tectonism, however.

Separation of the crustal and mantle components of surface heat flow is necessary to examine the contribution of crustal radiogenic heat production to the surface heat flow. This separation can only be attempted in the seventeen heat flow provinces which have been defined with surface heat flow and heat production data (see above and Vitorello and Pollack, 1980). The heat flow and heat production characteristics of these provinces are listed in Table 1. Also listed in Table 1 are the estimated crustal components of heat flow for each province, calculated by subtracting the reduced heat flow values from the heat flow/heat production regression analyses from the mean surface heat flow values. For the examination of the radiogenic component of surface heat flow, each heat flow province has been assigned an age based on the basement age in the province, presumably when the crust in that area assumed its primary heat production characteristics. The crustal component of heat flow is plotted as a function of this age in Figure 17.

No simple relationship between the crustal component of heat flow and crustal age is apparent from the data plotted in Figure 17. There are large error bars on the data from Phanerozoic provinces which represent in part transient thermal disturbances in the crust. However, the mean crustal component of heat flow for the Phanerozoic provinces, 24 ± 8 mW m^{-2} (n = 6), is not significantly different from the mean for Proterozoic provinces, 27 ± 15 mW m^{-2} (n = 7). An argument could be made that the data from the Early Proterozoic province of Central Australia should be ignored as divergent data, but similar data have also been measured in the Midcontinent of the United States (R.F. Roy, personal communication, 1982; D.D. Blackwell, personal communication, 1983), and if the Central Australian province is excluded, the mean crustal component of heat flow for Proterozoic provinces, 23 ± 9 mW m^{-2} (n=6), is even closer to the Phanerozoic mean. Thus, from the available data, there is no significant reduction of the crustal component of heat flow, as defined above, by erosion of

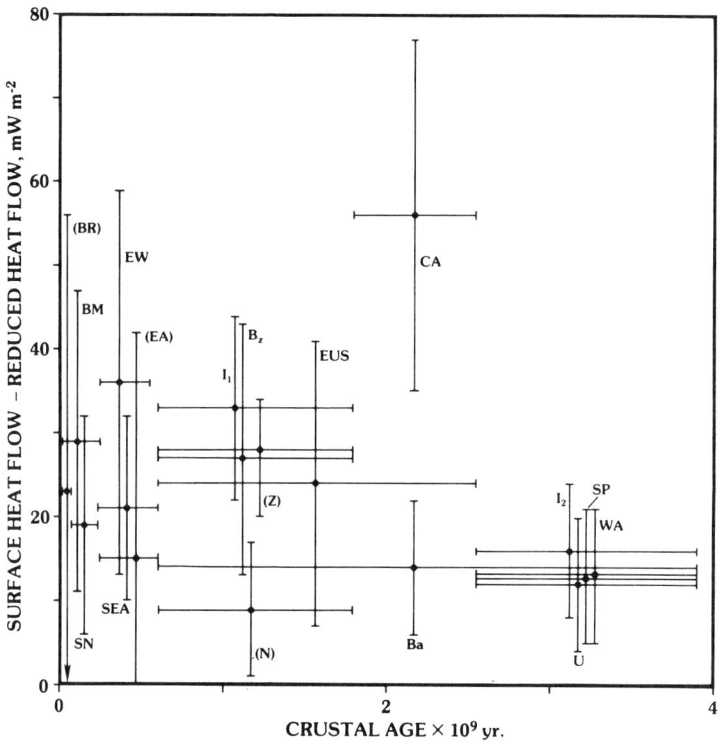

Figure 17. Crustal heat production q_A (defined by mean surface heat flow minus reduced heat flow) as a function of crustal age, for the heat flow provinces listed in Table 1. See Table 1 for province codes and age assignments. Heat flow error bars are plotted according to the standard deviation of the mean surface heat flow, and data are displaced as required on the age axis to clarify the plot. Data derived from assumed heat production depth parameters are indicated by codes in parentheses.

crustal heat producing elements at Phanerozoic and Paleozoic sites.

The crustal component of heat flow from Archean provinces, 14 ± 2 mW m^{-2} (n = 4) does appear to be different from the crustal component from younger provinces, being both smaller in magnitude, and more uniform than the other data sets. This difference and small scatter in the data are probably responsible for the lower heat flow and smaller scatter in the heat flow data from Archean sites in the global heat flow data set compared to data from younger sites (Figure 9). One argument is that the relative uniformity and low mean level of crustal heat production at Archean sites indicates that most of the original variability in near surface heat production has been eroded away at these sites. This simple explanation does not explain the apparent change in mean crustal heat production between Archean and Proterozoic sites, however. An increase in the mean ratio of the oxides of potassium to sodium has been shown to occur at the Archean/Proterozoic boundary (Engel et al., 1974), which suggests that the heat production change is the result of a basic chemical change in crustal composition at the end of the Archean. This change in basic chemical composition has been linked to a different style of tectonics and crustal evolution in the Archean (e.g., Engel et al., op cit; Fyfe, 1978; Morgan, 1983b). In conclusion, from the data shown in Figure 17, the crustal component of heat flow does not indicate that crustal radioactivity generally decreases with increasing age of tectonism through erosion, although heat production at Archean sites may be genetically lower.

If the data discussed above do not indicate a decrease in crustal heat production with time since tectonism, why erosion has not significantly reduced crustal heat production in older terrains must somehow be explained. One possible explanation of this apparent general uniformity with age is that older terrains were enriched with heat producing elements relative to younger terrains, and that erosion has equalized this initial imbalance. Some evidence exists for enrichment in older rocks, such as high crust heat production from Central Australia (Figure 17), but more data are required to test this hypothesis. A second possible explanation of the apparent uniformity of heat production is that heat producing elements in the crust remaining after erosion were never mobilized and concentrated upwards, and hence were never eroded, or were replaced. To investigate this second hypothesis, we must examine the processes that cause redistribution of heat producing elements in areas that are likely to be the sites of deep erosion. The most likely sites for deep erosion are zones of crustal thickening which occur along convergent plate boundaries.

In continental arcs, crustal thickening appears to be primarily due to magmatic activity in the volcanic arc of the system. High heat flow and the ascent of magmas will aid in the upward concentration of uranium and thorium in these zones, and thus it would be expected that subsequent erosion would remove much of the crustal radioactivity. However, magmas possibly add new heat producing elements to these zones, which compensate for radionuclides subsequently lost by erosion. The recent discovery of the occurrence of the relatively short lived isotope ^{10}Be in island arc lavas has been interpreted to indicate that material that was once on the ocean floor, was carried down with the subducted plate, and ejected in the arc lavas (Brown et al., 1982), in support of the concept of massive sediment subduction (e.g., Gilluly, 1971; Sibley and Vogel, 1976; Fyfe, 1978). Presumably uranium and thorium in ocean floor sediments would also be injected with arc magmatic activity, and thus the convergence zone may act as a system for lateral concentration of heat producing elements into the thickened crust. Heat producing elements lost by erosion may be effectively replaced in magmas.

In continental collision zones, the vertical sum of crustal heat production is increased by underthrusting. If the crustal thickness is doubled by underthrusting, then the crustal heat production would be approximately doubled. However, after complete erosion of the upper plate of the thrust, the remaining crust will only be depleted in heat producing elements if they were mobilized and concentrated into the upper plate prior to the completion of erosion. Widespread magmatic activity would certainly cause the lower plate to be depleted in heat producing elements. However, without such magmatic activity, the degree of mobilization will depend on the metamorphic grade attained by the lower plate during burial. Uranium appears to be essentially immobile in metamorphic rocks up to amphibolite grade, but is lost through migrating fluid phases during granulite metamorphism (Dostal and Capedri, 1978). Thorium would probably have a similar behavior (McLennan and Taylor, 1980). Thus, the heat producing elements in the lower plate of a double thickened crust would only be lost into the upper plate if the upper level of the lower plate was completely metamorphosed to granulite facies, which indicates a temperature of 700-800 °C (e.g., Turner and Verhoogen, 1960, p. 557). Without widespread magmatic activity this temperature range is difficult to attain at mid-crustal levels (P. Morgan and L.D. Ashwal, in preparation), and thus, without extensive magmatism, possibly the lower plate may retain most of its heat producing elements. In a doubly thickened crust in which the heat producing elements were not mobilized and concentrating upwards, heat production in the lower plate would not be represented by surface heat production samples, and would be represented in a heat flow/heat production regression analysis as a component of the reduced heat flow.

In summary, radiogenic heat production in the enriched zone of the upper crust is an important, but highly variable component of surface heat flow. For Phanerozoic and Proterozoic heat flow provinces the available data do not require a decreasing contribution to surface heat flow from the enriched zone with increasing crustal age, and the variability of this component does not appear to diminish with increasing crustal age. Only Archean sites show a consistent low crustal radiogenic component of heat flow, and there is remarkably little scatter in the means of data from these sites. The effects of removal of crustal heat producing elements by erosion may be somewhat counteracted by processes which redistribute the heat producing elements in convergence zones, the regions in which erosion is most likely to be significant. Low crustal heat production in Archean terrains is probably a genetic rather than an erosional effect, and there is no strong evidence that the crustal radiogenic heat production component of surface heat flow decreases with crustal age.

C. Equilibrium Mantle Heat Flux

Before considering the component of surface heat flow due to tectonic and magmatic activity, data relevant to the equilibrium mantle heat flux will be examined as it will be shown that these data constrain the maximum thickness of the continental lithosphere. The most pertinent data to the mantle heat flux are reduced heat flow values, and these data are listed in Table 1. As mantle heat flux is obviously more dependent on the tectonic and magmatic activity than the age of the overlying crust, reduced heat flow values are plotted as a function of the age of the last tectonothermal event in Figure 18. These data indicate that reduced heat flow is very variable, but generally high in regions of Cenozoic tectonothermal activity, is essentially undefined in regions of Mesozoic activity, but appears relatively constant for areas of Paleozoic and Precambrian activity. The most divergent data from areas of pre-Mesozoic activity are the data from Zambia (Z), the Late Proterozoic Indian Shield (I_1) and the Niger (N). A suggestion has been made that the Zambia data are perturbed by the effects of incipient rifting (Chapman and Pollack, 1974), but also possible is that they have a Mesozoic thermal component associated with Karroo volcanism in Africa. A Cenozoic or Mesozoic age may thus be more appropriate for these data (Z'). Similarly it may be possible that the Late Proterozoic Indian Shield data include a transient component associated with the Cretaceaous to Paleocene volcanism which produced the Deccan Trap Series (I_1'). Data from the Niger are based on very low surface heat flow values which may not be regionally representative (cf. Chapman and Pollack, op cit, Beck and Mustonen, 1972, and Sass and Behrendt, 1980). If all pre-Mesozoic reduced heat flow values are averaged and the uncorrected Superior province value is used, a mean of 27 ± 8 mW m^{-2} (n=13) is obtained. If the Zambia, Late Proterozoic Indian, and Niger data are excluded, and the corrected Superior value is used, the mean reduced heat flow from pre-Mesozoic sites is unchanged, but the standard deviation is reduced, 27 ± 4 mW m^{-2} (n=10).

Until more high quality reduced heat flow data are available, determining whether the variations in reduced heat flow values among different stable provinces are

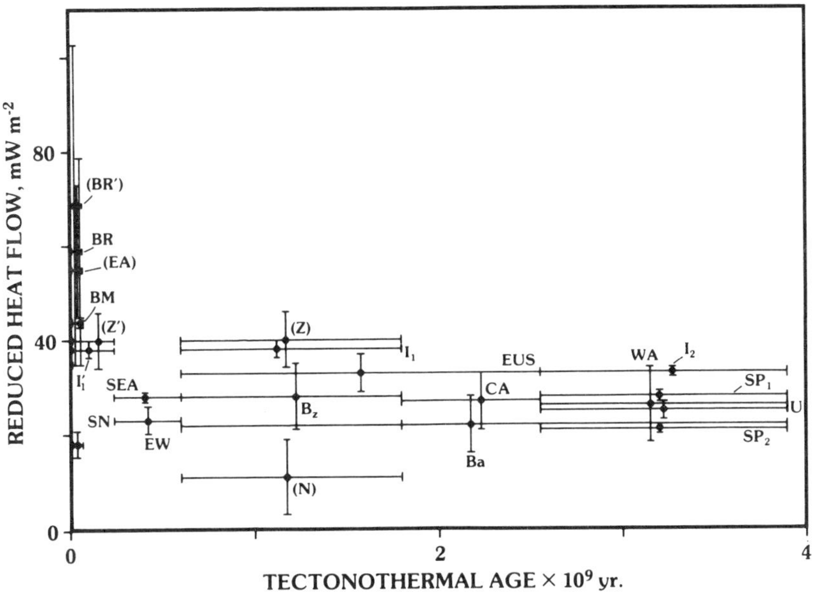

Figure 18. Reduced heat flow q* as a function of age since last tectonothermal event, for the heat flow provinces listed in Table 1. See Table 1 and text for province codes and age assignments. Heat flow error bars are uncertainties in reduced heat flow, and data are displaced as required on the age axis to clarify the plot. Data derived from assumed heat production depth parameters are indicated by codes in parentheses.

significant, or beyond the resolution of the data is difficult. Even if these variations are real, determining whether the variations are due to variations in heat input from the asthenosphere, possibly complicated by temporal variations in this thermal component, or due to crustal and/or mantle radiogenic heat production not removed by the heat flow/heat production regression analysis (e.g., see Allis, 1979) is impossible. These problems remain unanswered at present, but the mean reduced heat flux can be used to estimate typical mantle temperature gradients in a stable thermal regime.

Using the empirical relationship of Pollack and Chapman (1977) between surface and reduced heat flow, a reduced heat flow of 27 mW m^{-2} would typically be represented by a surface heat flow of 45 mW m^{-2}, the component of heat flow due to upper crustal radioactivity being 18 mW m^{-2} (equivalent to a surface heat production, A_s, of 1.8 μW m^{-3} with a depth distribution parameter, b, of 10 km). Assuming that the upper crustal radioactivity is distributed in a manner intermediate between the step and exponential distributions, that the depth parameter, b, appropriate to these distributions is 10 km, and that the upper crustal thermal conductivity is 2.5 W m^{-1} K^{-1}, the upper crustal radioactivity would cause an increase in temperature at depth of approximately 100 ^{o}C. If we may assume that the reduced heat flow is completely derived from conducted heat flow from the asthenosphere, and using a mantle conductivity of 3.4 W m^{-1} K^{-1} (see above), then the mean geothermal gradient in the mantle lithosphere is 7.9 ^{o}C km^{-1}. Alternatively, assuming that a portion of the reduced heat flow is derived from lower crustal radioactivity, and taking this component to be 7.5 mW m^{-2}, the middle of the range suggested for this component by Allis (1979), the corresponding mantle geothermal gradient is reduced to 5.7 ^{o}C km^{-1}, and the temperature at depth is increased a further 30 ^{o}C due to lower crustal radioactivity.

Defining a single temperature for the base of the thermal lithosphere is impossible, but by considering the processes which cause the transition for conductive to convective heat transfer, a range of temperatures for this transition can be estimated. For ultramafic rocks, and conditions relevant to the base of the lithosphere it is estimated that the transition from conductive to convective heat transfer occurs over a range of flow velocities on the order of 10^{-8} to 10^{-10} m s^{-1} ($3x10^{-1}$ to $3x10^{-3}$ m $year^{-1}$). These velocities are similar to velocities implied in the upper asthenosphere from glacial rebound data (e.g., Peltier _et al._, 1981). The effective viscosity of the asthenosphere has been estimated to be on the order of 10^{21} Pa s(10^{22} poise) from glacial rebound data (Peltier _et al._, _op cit_), and thus the base of the lithosphere can be taken as the depth at which the effective viscosity of the mantle decreases to on the order of 10^{21} Pa s. From creep deformation data and theory (e.g., Weertman, 1970), the effective viscosity of ultramafic rocks is predicted to decrease to 10^{21} Pa s at temperatures typically in the range 1300 to 1400 ^{o}C (e.g., Chapman and Pollack, 1974). Thus, the typical temperature range in which the transition from conductive to convective heat transfer, and thus the base of the lithosphere occurs, is estimated to be 1300 to 1400 ^{o}C.

Using the temperature gradients estimated above for the whole lithosphere, accounting for the temperature rise due to crustal radioactivity, and using a temperature range of 1300 to 1400 ^{o}C for the base of the lithosphere, the stable thickness of the continental lithosphere is estimated to be within the range 150 to 220 km. If a low conductivity crust is included in the analysis, both limits to the lithospheric thickness are reduced slightly to within the range 140 to 210 km. Geotherms corresponding to the range of lithospheric thickness predicted by these calculations are shown in Figure 19. These geotherms are reasonably consistent with geotherms estimated from the petrology of xenoliths in kimberlite pipes in South Africa (e.g., Harte, 1978).

Many variations on the models for the thermal structure of the continental lithosphere are possible, in addition to those presented above. With the constraints on thermal conductivity, radiogenic heat production, and vertical distribution of heat production imposed by the various data sets discussed above, however, only models with improbable parameters yield lithospheric thicknesses significantly outside the range predicted above. Thus, the mean thickness of the stable continental lithosphere is unlikely to exceed 140 to 210 km.

Variations in the crustal heat production component of heat flow, as indicated in Figure 19, will cause variations in the stable lithospheric thickness. For example, low heat production in Archean crust may allow the lithosphere to be 5 to 10 km thicker, and may be one factor in the stability of Archean cratons; high heat production in the Central Australian Shield and an exponential heat production distribution would constrain the lithosphere to be on average 20 to 35 km thinner, and very high heat production (A_s = 5.0 μW m^{-3}, b = 15 km) could make the lithosphere 35 to 50 km thinner. Thus, if regional variations in crustal heat production are included in the analysis, then the thickness range for the stable continental lithosphere is increased to be approximately 90 to 220 km.

D. Transient Heat Flow from Tectonogenesis

The thermal structure of the continental lithosphere can be modified in a number of ways during tectonism. The form of this modification will depend on the style of tectonism, and may even vary temporally and spatially in a single tectonic event. The magnitude of thermal modification will depend on the duration and intensity of tectonism. Most tectonic thermal perturbations to the lithosphere tend to increase the surface and reduced heat flows, but where magmatism is absent, a reduction in heat flow may be associated with plate convergence, such as along the western margin of portions of North and South America discussed above, and in the Sierra Nevada of North America (Blackwell, 1971; Roy *et al.*, 1972). The effects of erosion and sedimentation may continue to perturb the thermal structure of the continental lithosphere after the cessation of tectonism and magmatism, and may delay thermal relaxation of the lithosphere.

The recovery of surface heat flow from transient tectonic and magmatic disturbances will depend initially upon the depth of the main thermal perturbation in the lithosphere (e.g., Vitorello and Pollack, 1980). During major tectonothermal events, however, the complete thickness of the lithosphere will be thermally perturbed, in fact the lithosphere will probably be thickened or thinned. The time constant for thermal relaxation of the lithosphere will then depend primarily upon the equilibrium thickness of the continental lithosphere. As argued above, this thickness is on the order of 90 to 220 km, and thus the thermal time constant is constrained to be on the order of 100 to 500 m.y. (e.g., Oldenburg, 1975). This estimate for the thermal time constant for the continental lithosphere is consistent with the apparent uniformity in reduced heat flow for provinces stable since the Paleozoic shown in Figure 18.

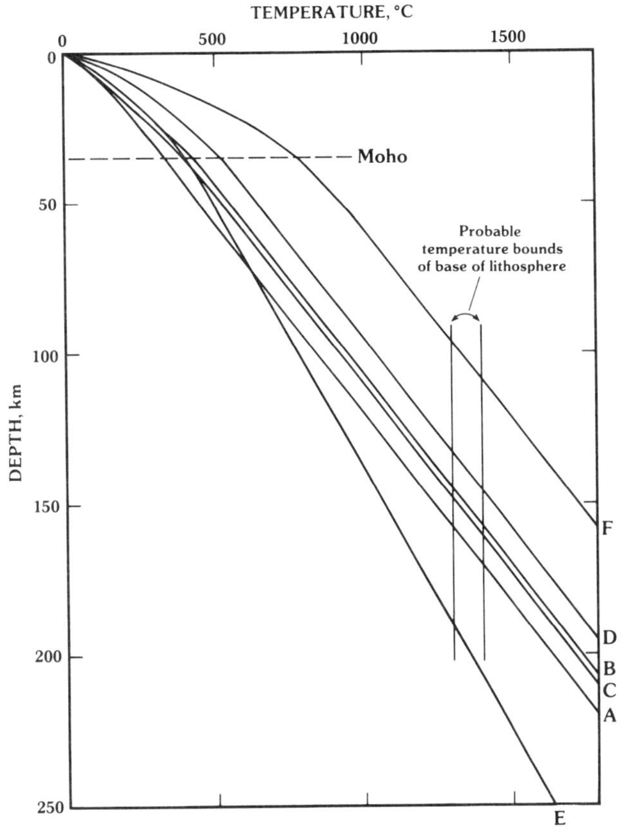

Figure 19. Calculated geotherms for the stable continental lithosphere. All geotherms calculated for a reduced heat flow q* of 27 mW m^{-2}, mantle thermal conductivity of 3.4 W m^{-1} K^{-1}, and upper crustal heat production to be the average of the step and exponential models. For geotherm A), surface heat production A_s = 1.8 μW m^{-3}, depth distribution parameter b = 10 km; B) as A) with 35 km thick crust with conductivity of 2.5 W m^{-1} K^{-1}, C) as B) with A_s = 1.0 μW m^{-3}; D) as B) with A_s + 5.0 μW m^{-3}; E) as B) with additional heat production totalling 7.5 mW m^{-2} between 20 and 35 km; and F) as B) with A_s = 5.0 μW m^{-3}, b = 15 km and exponential heat production distribution assumed.

Blackwell and Chockalingam (1980), using reduced heat flow data for North America, have deduced a similar time constant of 150±50 m.y. from these data, and estimated a thickness of the thermal lithosphere of 150±25 km.

As there are many forms to the thermal perturbation in the lithosphere during tectonism and magmatism, defining a single path along which the reduced heat flow will follow during and subsequent to tectonothermal activity is impossible. Each tectonothermal event is probably somewhat different, and thus the reduced heat flow history will be

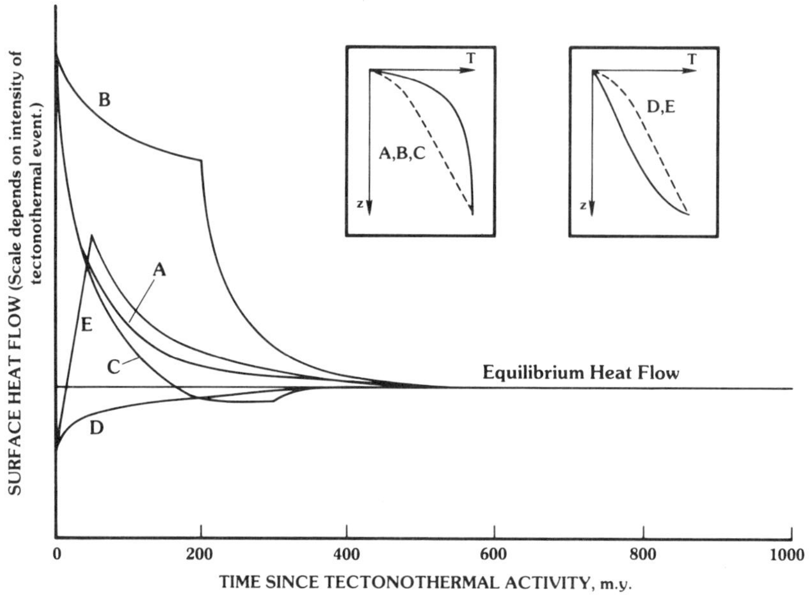

Figure 20. Qualitative surface heat flow histories following a variety of tectonomagmatic thermal perturbations. Curve A), initial high heat flow, no erosion or sedimentation; B), initial high heat flow, erosion for 200 m.y.; C), initial high heat flow, sedimentation for 300 m.y.; D) initial low heat flow, no erosion or sedimentation; and E) initial low heat flow, rapid erosion for 50 m.y. The shape of the initially disturbed geotherm for each curve is given in the inset boxes with reference to stable geotherm (dashed). The scale for the heat flow axis depends on the intensity of the initial disturbance. The time axis is appropriate for an equilibrium lithosphere thickness on the order of 150 km.

different for each event. Semiquantitative estimates of the reduced heat flow histories for a number of different events are shown in Figure 20. As shown, erosion or sedimentation may prolong the thermal perturbation. A conclusion can be made from these results that the effects of tectonogenesis on surface heat flow probably last no more than a few 100 m.y.

E. The Secular Variation of Surface Heat Flow

The global heat flow data set of Chapman and Furlong (1977), plotted as a function of age of the last tectonothermal event at each heat flow site is shown in Figure 8, and was discussed briefly above. Two possible interpretations of this data set were given; the first that the data indicate a continuous secular decrease of heat flow with age of tectogenesis (modelled by Vitorello and Pollack,

1980), and the second that there is a large variation in heat flow at Cenozoic, Mesozoic, and possibly Late Paleozoic sites, heat flow is constant at sites ranging in age from Early Paleozoic through Proterozoic, and slighlty lower at Archean sites. The data and arguments presented above support the second interpretation of the secular variation of heat flow.

Highly variable, and generally high heat flow in Cenozoic and Mesozoic tectonothermal terrains is the result of transient thermal perturbations caused by tectonic and magmatic activity in these terrains, and the scatter in the data is increased by the strong effects of the shallow thermal regime in these young terrains. In late Paleozoic terrains only a small thermal anomaly remains, the duration of which may have been extended by the effects of erosion. Heat flow data from Early Paleozoic and Proterozoic terrains represent stable continental lithosphere with possibly a small, but unresolvable variation in heat input from the asthenosphere, but most of the variation in surface heat flow resulting from variations in upper crustal radiogenic heat production. A decrease in this crustal component of heat flow with crustal age due to erosion is not required by the data. Heat flow in Archean terrains is basically similar to that in Proterozoic terrains, but the contribution from crustal heat production appears to be lower and more uniform, resulting in slightly lower and more uniform surface heat flow.

8. CONCLUDING REMARKS

Simple models have been relatively successful at predicting the thermal structure and evolution of oceanic lithosphere (e.g., Oldenburg, 1975; Crough and Thompson, 1976a; Parsons and Sclater, 1977; Sclater et al., 1980; Von Herzen et al., 1982). The continental heat flow data set appears to express a more complex thermal structure and evolution, however. Variations in quantity and distribution of heat production and the continental lithosphere, primarily reflecting variations in the chemical evolution, result in variations in the equilibrium thermal structure and surface heat flow for continental lithosphere. The time constant for the continental lithosphere appears to be on the order of a few 100 m.y., similar to the oceanic lithosphere time constant, but thermal relaxation of continental lithosphere can be delayed by the effects of post-tectonic sedimentation and erosion. Tectonism and magmatism in the continental lithosphere can result in a variety of thermal perturbations, and thus a single thermal model cannot be applied to the thermal evolution of the lithosphere in different tectonic environments. The global heat flow data set averages real scatter in the data caused by near surface thermal effects, multi-modal local data sets caused by variations in crustal heat production, and a variety of thermal relaxation patterns resulting from a variety of initial thermal perturbations. More data, especially complementary heat flow/heat production data sets, are required to further unravel the complex thermal structure and evolution of the continental lithosphere.

ACKNOWLEDGEMENTS

As with any review article, this paper has drawn heavily on the work of others, particularly the collectors and compilers of the continental heat flow data set. I acknowledge all my colleagues in the heat flow community, particularly J. Wheildon, who introduced me to heat flow studies, D.D. Blackwell, who has tutored me in all aspects of terrestrial heat flow over the past decade, and A.H. Lachenbruch, H.N. Pollack and J.H. Sass, who have shared many of their ideas on continental heat flow with me, and patiently listened to my own thoughts. My colleagues at the Lunar and Planetary Institute have added depth to many of the tectonic and petrologic discussions in this paper. However, I remain responsible for any errors and omissions.

M.K. Morgan and J.H. Sass are thanked for reviews of a draft of this manuscript. S. Adlis, D. Chady, B. Fessler, K. Hrametz, K. Leung, and P. Thompson are thanked for assistance in the final preparation of this manuscript. This study was carried out at the Lunar and Planetary Institute, which is operated by the Universities Space Research Association under contract no. NASW-3389 from the National Aeronautics and Space Administration. This is LPI contribution number 520.

REFERENCES

Allis, R.G. (1978) The effect of Pleistocene climatic variations on the geothermal regime in Ontario: a reassessment. Can. J. Earth Sci. 15, 1875-1879.

Allis, R.G. (1979) A heat production model for stable continental crust. Tectonophysics 57, 151-165.

Anderson R.N. (1980) 1980 update of heat flow in the east and southeast Asian seas. In The Tectonic and Geologic Evolution of Southeastern Asian Seas and Islands (Hayes, D.E., ed) American Geophysical Union, Washington, D.C., pp. 318-326.

Atkinson, B.K. (1982) Shear stress in fault zones. Nature 296, 708.

Baker, B.H. and Morgan, P. (1981) Continental rifting: Progress and outlook. Eos, Trans. Am. Geophys. Un. 62, 585-586.

Baker, B.H., Mohr, P.A. and Williams, L.A.J. (1972) Geology of the eastern rift system of Africa. Geol. Soc. Am. Spec. Pap. 136, 67 pp.

Barrell, J. (1914a) The strength of the Earth's crust, 5. J. Geol. 22, 441-468.

Barrell, J. (1914b) The strength of the Earth's crust, 6. J. Geol. 22, 655-683.

Beck, A.E. (1965) Techniques of measuring heat flow on land. In Terrestrial Heat Flow (Lee, W.H.K., ed) American Geophysical Union, Washington, D.C., pp. 24-57.

Beck, A.E. and Mustonen, E. (1972) Preliminary heat flow data from Ghana. Nature Phy. Sci. 235, 172-174.

Benfield, A.E. (1949) The effect of uplift and denudation on underground temperatures. J. Appl. Phys. 20, 66-70.

Bickle, M.J., Hawkesworth, C.J., England, P.C. and Athey, D.R. (1975) A preliminary thermal model for regional metamorphism in the Eastern Alps. Earth Planet. Sci. Lett. 26, 13-28.

Birch, F. (1948) The effects of Pleistocene climatic variations upon geothermal gradients. Am. J. Sci. 246, 729-760.

Birch, F. (1950) Flow of heat in the Front Range, Colorado. Bull. Geol. Soc. Am. 61, 567-630.

Birch, F.S. (1975) Conductive heat flow anomalies over a hot spot in a moving medium. J. Geophys. Res. 80, 4825-4827.

Birch, F. and Clark, H. (1940) The thermal conductivity of rocks and its dependence upon temperature and composition. Am. J. Sci. 238, 613-635.

Birch, F., Roy, R.F. and Decker, E.R. (1968) Heat flow and thermal history in New York and New England. In Studies of Appalachian Geology: Northern and Maritime (Zen, E., White, W.S., Hadley, J.B. and Thompson, J.B., Jr., eds) Interscience, New York, pp. 437-451.

Bird, P. (1978a) Initiation of intracontinental subduction in the Himalaya. J. Geophys. Res. 83, 4975-4987.

Bird, P. (1978b) Finite element modeling of lithospheric deformation: The Zagros collision orogeny. Tectonophysics, 50, 307-336.

Bird, P., Toksöz, M.N. and Sleep, N.H. (1975) Thermal and mechanical models of continent-continent convergence zones. J. Geophys. Res. 80, 4405-4416.

Blackwell, D.D. (1971) The thermal structure of the continental crust. In The Structure and Physical Properties of the Earth's Crust (Heacock, J.G., ed) American Geophysical Union, Washington, D.C., pp. 169-184.

Blackwell, D.D. (1978) Heat flow and energy loss in the western United States. In Cenozoic Tectonics and Regional Geophysics of the Western Cordillera (Smith, R.B. and Eaton, G.P., eds) Geological Society of America, Memoir 152, Boulder, Colorado, pp. 175-208.

Blackwell, D.D. and Chockalingam, S. (1980) Thermal evolution of the continental lithosphere. Abs. International Conference on Mathematical Problems of the Thermal and Dynamic State of the Earth, July 28- August 3, 1980, Lake Arrowhead, California, pp. 105-108.

Blackwell, D.D., Steele, J.L. and Brott, C.A. (1980) The terrain effect on terrestrial heat flow. J. Geophys. Res. 85, 4757-4772.

Blackwell, D.D., Bowen, R.G., Hull, D.A., Riccio, J. and Steele, J.L. (1982) Heat flow, arc volcanism, and subduction in northern Oregon. J. Geophys. Res. 87, 8735-8754.

Bridwell, R.J. and Potzick, C. (1981) Thermal regimes, mantle diapirs and crustal stresses of continental rifts. Tectonophysics 73, 15-32.

Brott, C.A., Blackwell, D.D. and Ziagos, J.P. (1981) Thermal and tectonic implications of heat flow in the eastern Snake River Plain, Idaho. J. Geophys. Res. 86, 11709-11734.

Brown, L., Klein, J., Middleton, R., Sacks, I.S. and Tera, F. (1982) ^{10}Be in island-arc volcanoes and implications for subduction. Nature 299, 718-720.

Brune, J.N., Henyey, T.L. and Roy, R.F. (1969) Heat flow, stress and rate of slip along the San Andreas fault. J. Geophys. Res. 74, 3821-3827.

Bucher, G.J. and Decker, E.R. (1980) Heat flow and heat production in the Ross Island-Dry Valley area, southern Victoria Land, Antarctica (abs). Eos, Trans. Am. Geophys. Un. 61, 1131.

Carvalho, H. da S., Purwoko, Siswoyo, Tharmin, M. and Vacquier, V. (1980) Terrestrial heat flow in the